Collins

The Shang[hai] Maths Project

For the English National Curriculum

Practice Book 2A

Series Editor: Professor Lianghuo Fan

UK Curriculum Consultant: Paul Broadbent

HarperCollins
PUBLISHERS
Since 1817

William Collins' dream of knowledge for all began with the publication of his first book in 1819.

A self-educated mill worker, he not only enriched millions of lives, but also founded a flourishing publishing house. Today, staying true to this spirit, Collins books are packed with inspiration, innovation and practical expertise. They place you at the centre of a world of possibility and give you exactly what you need to explore it.

Collins. Freedom to teach.

Published by Collins, an imprint of HarperCollins*Publishers*
The News Building
1 London Bridge Street
London
SE1 9GF

HarperCollins*Publishers*
Macken House, 39/40 Mayor Street Upper
Dublin 1, D01 C9W8, Ireland

Browse the complete Collins catalogue at
www.collins.co.uk

10 9 8 7 6

978-0-00-822609-1

Translated by Professor Lianghuo Fan, Adapted by Professor Lianghuo Fan.

British Library Cataloguing in Publication Data

A catalogue record for this publication is available from the British Library

Series Editor: Professor Lianghuo Fan
Uk Curriculum Consultant: Paul Broadbent
Publishing Manager: Fiona McGlade
In-house Editor: Nina Smith
In-house Editorial Assistant: August Stevens
Project Manager: Emily Hooton
Copy Editor: Catherine Dakin
Proofreaders: Karen Williams, Cassandra Fox and Gerard Delaney
Cover design: Kevin Robbins and East China Normal University Press Ltd
Cover artwork: Daniela Geremia
Internal design: 2Hoots Publishing Services Ltd
Typesetting: 2Hoots Publishing Services Ltd
Illustrations: QBS
Production: Rachel Weaver
Printed and Bound in the UK using 100% Renewable Electricity at CPI Group (UK) Ltd

The Shanghai Maths Project (for the English National Curriculum) is a collaborative effort between HarperCollins, East China Normal University Press Ltd. and Professor Lianghuo Fan and his team. Based on the latest edition of the award-winning series of learning resource books, *One Lesson, One Exercise*, by East China Normal University Press Ltd. in Chinese, the series of Practice Books is published by HarperCollins after adaptation following the English National Curriculum.

Practice Book Year 2A has been translated and developed by Professor Lianghuo Fan with the assistance of Ellen Chen, Ming Ni, Huiping Xu and Dr Lionel Pereira-Mendoza, with Paul Broadbent as UK Curriculum Consultant.

MIX
Paper | Supporting
responsible forestry
FSC™ C007454

This book is produced from independently certified FSC™ paper to ensure responsible forest management.
For more information visit: www.harpercollins.co.uk/green

Contents

1.1 Equal grouping

 Learning objective Make totals by equal grouping

 Basic questions

1 Circle the dots to group them and then write the missing numbers in the boxes. The first one has been done for you.

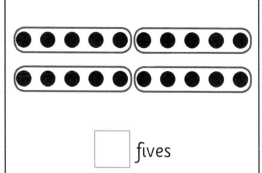

2 tens

☐ fives

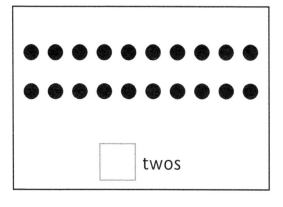

☐ twos

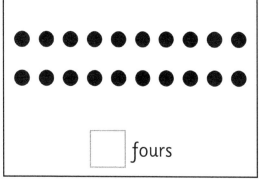

☐ fours

2 Draw circles to group.

3 fives	8 twos	6 threes	5 fours
2 fives	5 twos	4 threes	3 fours

3 Fill in the boxes. The first one has been done for you.

4	threes

4 Draw circles to group the sets of stars in different ways. Then fill in the boxes.

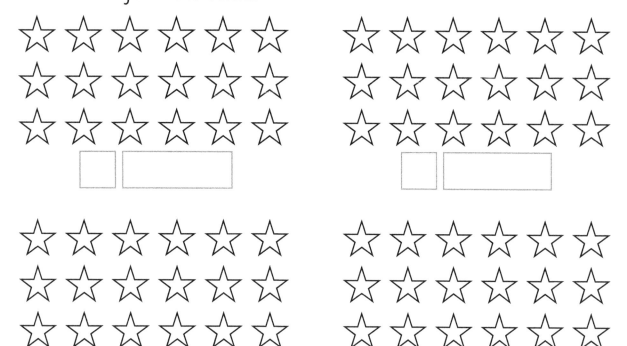

☆ ☆ ☆ ☆ ☆ ☆
☆ ☆ ☆ ☆ ☆ ☆
☆ ☆ ☆ ☆ ☆ ☆
☐ ☐

☆ ☆ ☆ ☆ ☆ ☆
☆ ☆ ☆ ☆ ☆ ☆
☆ ☆ ☆ ☆ ☆ ☆
☐ ☐

☆ ☆ ☆ ☆ ☆ ☆
☆ ☆ ☆ ☆ ☆ ☆
☆ ☆ ☆ ☆ ☆ ☆
☐ ☐

☆ ☆ ☆ ☆ ☆ ☆
☆ ☆ ☆ ☆ ☆ ☆
☆ ☆ ☆ ☆ ☆ ☆
☐ ☐

5 Draw the dots and then write the numbers in the spaces.

2 threes equals _____ .	●●● ●●●
4 twos equals _____ .	●● ○○ ○○ ○○
3 threes equals _____ .	
2 _____ equals 8.	
3 _____ equals 6.	
4 _____ equals 12.	

Challenge and extension question

6 Work out each calculation. Write your answers in the boxes.

(a) **|** = 3

|||| [ ] [ ] make [ ]

||||||| [ ] [ ] make [ ]

||||| [ ] [ ] make [ ]

(b) **■** = 5

■■■■ [ ] [ ] make [ ]

■■■■■ [ ] [ ] make [ ]

■■■■■■■■ [ ] [ ] make [ ]

1.2 The 100 square (I)

 Learning objective Explore patterns on a 100 square

 Basic questions

1 Count from 1 to follow these instructions. What do you notice?

(a) Count in threes and colour the squares that you land on.

1	2	3	4	5	6	7	8	9	10
11	12	13	14	15	16	17	18	19	20
21	22	23	24	25	26	27	28	29	30
31	32	33	34	35	36	37	38	39	40
41	42	43	44	45	46	47	48	49	50
51	52	53	54	55	56	57	58	59	60
61	62	63	64	65	66	67	68	69	70
71	72	73	74	75	76	77	78	79	80
81	82	83	84	85	86	87	88	89	90
91	92	93	94	95	96	97	98	99	100

(b) Count in fours and colour the squares that you land on.

1	2	3	4	5	6	7	8	9	10
11	12	13	14	15	16	17	18	19	20
21	22	23	24	25	26	27	28	29	30
31	32	33	34	35	36	37	38	39	40
41	42	43	44	45	46	47	48	49	50
51	52	53	54	55	56	57	58	59	60
61	62	63	64	65	66	67	68	69	70
71	72	73	74	75	76	77	78	79	80
81	82	83	84	85	86	87	88	89	90
91	92	93	94	95	96	97	98	99	100

2 Write the correct number in each ◯ on the 100 squares.

1		◯			◯				
11			◯		◯				
21				25					
31			◯		◯				◯
41		◯				◯			◯
51								59	◯
61								69	
71								◯	
81			◯	◯	◯	◯	◯	◯	
91			◯						

									10
			◯						20
			◯						30
	◯	◯	34	◯	◯				40
			◯					◯	50
			◯				◯		60
◯							◯		70
	◯				76				80
		◯		85					90
			◯						100

6 ◯ ◯ ◯ ◯ ◯ ◯ ◦

3 Let's make a beautiful design. Find and then colour the following numbers in the 100 square.

5, 6, 14, 17, 23, 25, 26, 28, 32, 34, 37, 39, 41, 43, 48, 50, 51, 53, 58, 60, 62, 64, 67, 69, 73, 75, 76, 78, 84, 87, 95, 96.

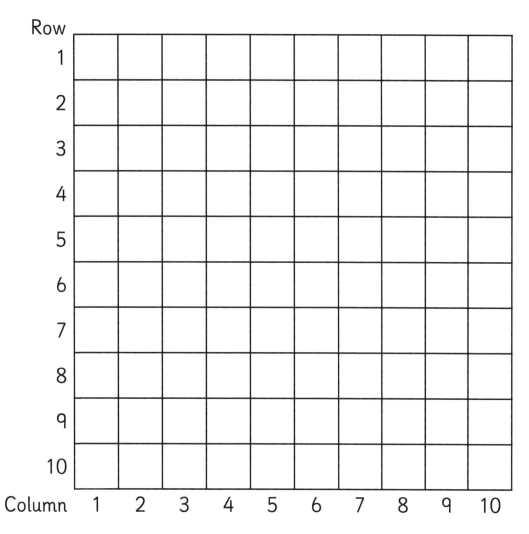

4 Use the 100 square to answer the questions.

			4	5	6	7			
		13					18		
	22							29	
31									40
			45	46					
			55	56					
61									70
	72							79	
		83					88		
			94	95	96	97			

(a) First, fill in the missing numbers in the 100 square in order.

(b) The number after 99 is ☐ .

(c) Look at the first nine numbers in the third row. The digits in the tens places are all ☐ .

(d) All the numbers in the sixth column have 6 in their

_____ places.

5 Use a 100 square to answer these questions.

(a) Write all the numbers with 9 in the tens places.

_____

Write all the numbers with 2 in the ones place.

_____

(b) Write all the whole tens greater than 30.

_____

Write the even numbers greater than 40 but less than 60.

_____

Write the odd numbers greater than 70 but less than 90.

_____

(c) Write every number that has the digit in the ones place 1 greater than the digit in the tens place.

_____

Write every number that has the digit in the tens place 2 greater than the digit in the ones place.

_____

1.3 Addition and subtraction of numbers up to 20

 Learning objective Add and subtract numbers within 20

 Basic questions

1 Calculate mentally.

8 + 3 = ☐	12 − 6 = ☐	13 − 8 = ☐	5 + 7 + 0 = ☐
15 − 2 = ☐	8 − 3 = ☐	0 + 12 = ☐	18 − 5 − 3 = ☐
10 − 9 = ☐	16 − 10 = ☐	17 − 7 = ☐	7 + 8 − 9 = ☐
6 + 5 = ☐	6 + 8 = ☐	19 − 12 = ☐	13 − 3 + 6 = ☐
10 + 7 = ☐	15 − 7 = ☐	11 − 0 = ☐	7 + 6 − 9 = ☐
20 − 10 = ☐	9 + 8 = ☐	20 − 14 = ☐	20 − 8 − 8 = ☐

2 Complete the number patterns.

3, ☐, 9, 12, ☐, ☐

18, 16, ☐, ☐, ☐, 8

☐, 15, 10, ☐, ☐

3 Use the three numbers in each group to make two addition sentences and two subtraction sentences.

    4 8 12   5 15 10   9 9 0

_____   _____   _____

_____   _____   _____

_____   _____   _____

_____   _____   _____

4 Complete the number walls.

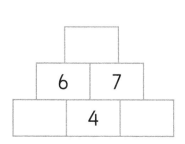
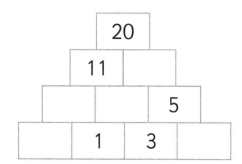

5 Calculate with reasoning.

6 + 3 =	17 − 8 =	16 − 9 =	3 + 11 =
16 + 3 =	15 − 8 =	16 − 8 =	4 + 10 =
6 + 13 =	13 − ☐ =	16 − ☐ =	5 + ☐ =
16 + 13 =	☐ − ☐ =	☐ − ☐ =	☐ + ☐ =

6 Without calculating, write >, < or = in each $\bigcirc$.

$9 + 6 \bigcirc 9 + 7$ $\qquad$ $18 - 9 \bigcirc 18 - 8$

$2 + 3 + 4 \bigcirc 2 + 4 + 5$ $\qquad$ $10 - 5 \bigcirc 10 - 6$

$4 + 16 \bigcirc 16 + 4$ $\qquad$ $6 + 9 + 3 \bigcirc 6 + 9 - 3$

$7 + 4 \bigcirc 8 + 3$ $\qquad$ $12 + 8 \bigcirc 15 + 5$

$12 - 8 - 3 \bigcirc 12 - 3 - 8$ $\qquad$ $12 - 5 \bigcirc 11 - 4$

$20 - 8 \bigcirc 16 - 8$ $\qquad$ $11 - 7 + 4 \bigcirc 11 + 4 - 7$

7 Write addition sentences with a sum of 16.

$16 = \square + \square = \square + \square = \square + \square = \square + \square$

$= \square + \square = \square + \square = \square + \square = \square + \square$

8 Write subtraction sentences with a difference of 9.

$9 = \square - \square = \square - \square = \square - \square = \square - \square$

$= \square - \square = \square - \square = \square - \square = \square - \square$

9 Use the numbers 5, 6, 7, 8, 9, 10, 11 and 12 to write addition and subtraction sentences.

Use each number only once.

Then work out the answers.

| 5 | 6 | 7 | 8 | 9 | 10 | 11 | 12 |

(a) ☐ + ☐ = ☐   (b) ☐ − ☐ = ☐

☐ + ☐ = ☐   ☐ − ☐ = ☐

☐ + ☐ = ☐   ☐ − ☐ = ☐

☐ + ☐ = ☐   ☐ − ☐ = ☐

1.4 Fun with calculation

 Learning objective Use the inverse relationship between addition and subtraction

 Basic questions

1 Calculate mentally.

5 + 4 = ☐	12 + 6 = ☐	0 + 17 = ☐	12 + 6 + 2 = ☐
9 – 3 = ☐	8 + 8 = ☐	20 – 8 = ☐	20 – 7 – 3 = ☐
7 – 7 = ☐	15 – 9 = ☐	7 + 5 = ☐	12 – 9 + 4 = ☐
19 – 8 = ☐	17 – 8 = ☐	14 + 6 = ☐	8 + 9 – 5 = ☐
9 + 3 = ☐	7 + 6 = ☐	15 – 7 = ☐	19 – 0 – 19 = ☐

2 Write + or – in each ◯.

13 ◯ 4 = 9   16 ◯ 2 = 14

4 ◯ 4 = 0   8 ◯ 4 ◯ 3 = 15

5 ◯ 10 = 15   8 ◯ 8 = 16

1 ◯ 12 = 13   8 ◯ 4 ◯ 3 = 1

10 ◯ 2 = 8   9 ◯ 3 = 12

19 ◯ 4 = 15   8 ◯ 4 ◯ 3 = 9

3 Draw the missing dots and write the correct numbers in the boxes.

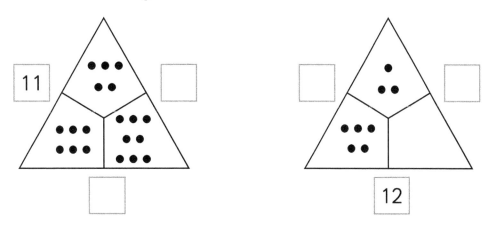

4 Find the missing numbers.

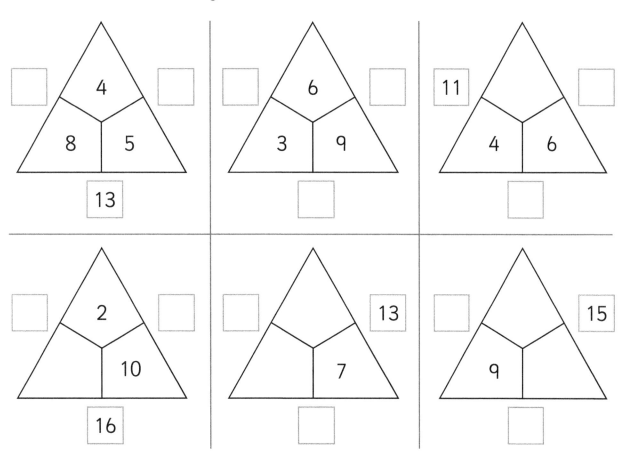

5 Fill in each ◯ with a number so that the sum of the three numbers on each line is 15.

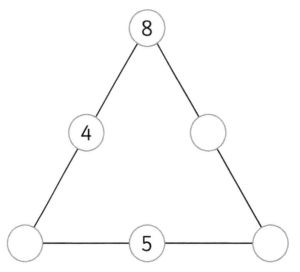

Challenge and extension question

6 Think carefully. Write the three missing numbers for each triangle.

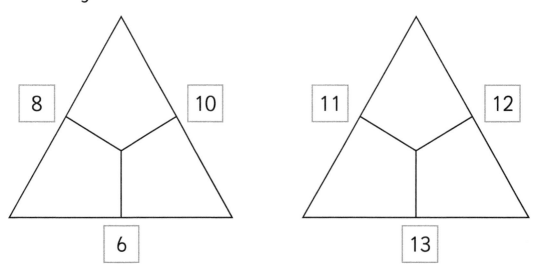

1.5 Comparing numbers

Learning objective Use < and > to compare addition and subtraction bonds

Basic questions

1 Calculate mentally.

$3 + 5 = \boxed{}$　　　$1 + 9 = \boxed{}$　　　$4 + 8 = \boxed{}$

$6 + 7 = \boxed{}$　　　$9 - 6 = \boxed{}$　　　$12 - 10 = \boxed{}$

$15 - 5 = \boxed{}$　　　$18 - 3 = \boxed{}$　　　$15 + 5 = \boxed{}$

$6 + 11 = \boxed{}$　　　$10 - 7 = \boxed{}$　　　$8 - 6 = \boxed{}$

$17 - 8 = \boxed{}$　　　$8 - 7 = \boxed{}$　　　$7 + 2 = \boxed{}$

$10 - 8 = \boxed{}$　　　$11 + 7 = \boxed{}$　　　$16 - 7 = \boxed{}$

2 Write >, < or = in each $\bigcirc$.

$9 - 3 \bigcirc 9$　　　$11 + 4 \bigcirc 15$　　　$14 + 4 \bigcirc 14 - 4$

$7 \bigcirc 6 + 3$　　　$6 - 6 \bigcirc 12$　　　$3 + 9 \bigcirc 5 + 7$

$16 \bigcirc 9 + 6$　　　$5 + 8 \bigcirc 18 - 6$　　　$12 - 10 \bigcirc 12 - 5 + 5$

3 Fill in the boxes using the numbers shown on the cards.

(a)

| 7 + ☐ < 13 | 7 + ☐ < 13 | 7 + ☐ < 13 |
| 7 + ☐ < 13 | 7 + ☐ < 13 | 7 + ☐ < 13 |

(b)

18 − ☐ < 8	18 − ☐ < 8	18 − ☐ < 8
18 − ☐ < 8	18 − ☐ < 8	18 − ☐ < 8
18 − ☐ < 8	18 − ☐ < 8	

4 What is the greatest number you can write in each box?

☐ + 8 < 12   11 − ☐ > 6   14 + ☐ < 20

15 − ☐ > 5   ☐ − 7 < 8   7 + ☐ < 12

20 − ☐ > 10   ☐ + 3 < 15   ☐ − 13 < 4

☐ 3 < 32   100 > ☐ 9   51 > 5 ☐

Challenge and extension questions

5 First row:

Second row:

I need to move ☐ 🍎 from the first row to the second so that both rows have the same number of apples.

6 Write >, < or = in each ◯.

$\triangle + 5 = \star + 3$ ⟶ $\triangle ◯ \star$

$\square + 4 = \triangle + 6$ ⟶ $\square ◯ \triangle$

$\diamond + 2 = ◎ - 2$ ⟶ $\diamond ◯ ◎$

Chapter 1 test

1 Calculate mentally.

$5 + 6 =$ ☐ $7 + 6 =$ ☐ $14 + 3 =$ ☐

$12 + 7 =$ ☐ $20 - 10 =$ ☐ $13 - 10 =$ ☐

$14 - 9 =$ ☐ $10 - 9 =$ ☐ $7 + 8 =$ ☐

$12 + 6 =$ ☐ $11 - 6 =$ ☐ $18 - 9 =$ ☐

$2 +$ ☐ $= 7$ ☐ $+ 9 = 15$ $14 -$ ☐ $= 7$

$8 +$ ☐ $= 19$ ☐ $- 9 = 8$ ☐ $- 6 = 8$

$1 + 4 + 9 =$ ☐ $5 + 2 + 8 =$ ☐ $9 + 4 - 7 =$ ☐

$15 - 6 - 4 =$ ☐   ☐ $+ 4 - 7 = 5$ ☐ $- 8 + 2 = 6$

2 Write the difference.

Minuend	10	17	14	15
Subtrahend	6	9	6	10
Difference				

Write the sum.

Addend	5	3	8	7
Addend	9	12	8	4
Sum				

3 Write >, < or = in each ◯ and write the correct symbol in the box.

14 − 7 ◯ 5   8 + 7 ◯ 9 + 4   12 − 4 ◯ 15 − 9

11 − 5 ◯ 7   3 + 7 ◯ 2 + 9   16 − 9 ◯ 14 − 6

20 ◯ 9 + 9   5 + 5 ◯ 12 − 8   12 ☐ 3 ◯ 3 + 6

4 Fill in the boxes.

6 + | 5 |
    | 6 |
    | 7 | < ☐
    | 8 |
    | 9 |

7 + | |
    | |
    | | < 17
    | |
    | |

☐ + | 5 |
    | 6 |
    | 7 | < 18
    | 8 |
    | 9 |

5 What is the greatest number you can write in each box?

3 + ☐ < 12   ☐ + 8 < 15   8 + ☐ < 18 − 1

15 − ☐ > 6   14 − ☐ > 7 + 4   9 + 2 > ☐ − 3

☐ 6 < 82   50 > ☐ 9   94 > 8 ☐

6 Read the number lines and then write the number sentences.

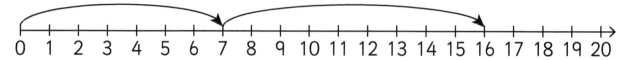

_____

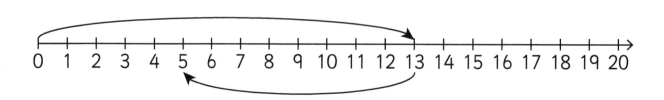

_____

7 Write the number sentences and then calculate.

   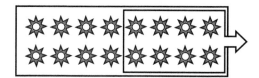

_____   _____

What is the difference?   How many rockets are there altogether?

_____   _____

(a) For each diagram, choose five numbers from the cards to fill in the ◯ so the sum of the three numbers on each line is the same.

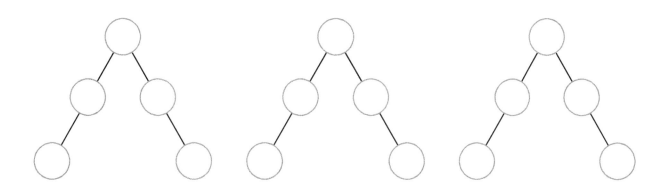

(b) What number does each shape stand for?

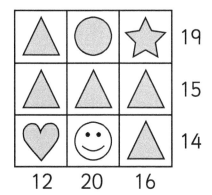

△	◯	☆	19
△	△	△	15
♥	☺	△	14
12	20	16	

△ = ☐	♥ = ☐
☆ = ☐	◯ = ☐
☺ = ☐	

Chapter 2 Addition and subtraction within 100

2.1 Adding or subtracting tens to or from 2-digit numbers

 Learning objective Add or subtract tens to or from 2-digit numbers

 Basic questions

1 Look at the diagrams and complete each calculation.
Write the missing numbers in the boxes.

Tens	Ones
● ● ●	○ ○
● ●	

32 + ☐ = ☐

Tens	Ones
⊘ ⊘ ⊘ ⊘ ● ● ● ●	○ ○ ○

83 − ☐ = ☐

2 Calculate mentally.

38 + 40 = ☐   30 + 49 = ☐   84 – 80 = ☐   26 + 70 = ☐

93 – 60 = ☐   12 + 70 = ☐   75 – 50 = ☐   48 – 30 = ☐

15 + 50 = ☐   61 – 20 = ☐   20 + 36 = ☐   74 – 60 = ☐

97 – 80 = ☐   85 – 70 = ☐   70 + 25 = ☐   100 – 80 = ☐

3 Write + or – in each ◯.

53 ◯ 40 = 93   56 ◯ 20 = 36   40 ◯ 24 = 64

87 ◯ 30 = 57   25 ◯ 10 = 15   86 ◯ 80 = 6

61 ◯ 20 = 41   18 ◯ 60 = 78   100 ◯ 40 = 60

4 Write the missing numbers in the boxes.

49 + ☐ = 69   75 – ☐ = 35   ☐ – 20 = 48

☐ + 24 = 84   ☐ – 30 = 12   85 – ☐ = 5

66 – ☐ = 16   100 – ☐ = 40   ☐ + 48 = 78

5 Write the missing numbers in the boxes.

(a) 16 $\xrightarrow{+\ 20}$ ☐ $\xrightarrow{+\ 20}$ ☐ $\xrightarrow{+\ 20}$ ☐ $\xrightarrow{+\ 20}$ ☐

(b) 95 $\xrightarrow{-\ 20}$ ☐ $\xrightarrow{-\ 20}$ ☐ $\xrightarrow{-\ 20}$ ☐ $\xrightarrow{-\ 20}$ ☐

(c) 42 $\xrightarrow{+\ 30}$ ☐ $\xrightarrow{-\ 40}$ ☐ $\xrightarrow{+\ 50}$ ☐ $\xrightarrow{-\ 60}$ ☐

Challenge and extension question

6 Fill in each circle with a whole ten so that the sum of the three numbers on each line is 100. Make each spider's web different.

2.2 Adding or subtracting a 1-digit number to or from a 2-digit number (1)

 Learning objective Add or subtract a 1-digit number to or from 2-digit numbers

 Basic questions

1 Use the number lines to write addition and subtraction calculations. Write the answers in the boxes.

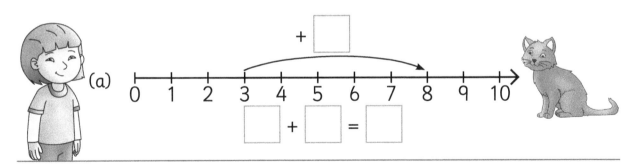

(a)

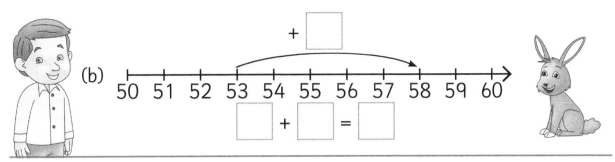

(b)

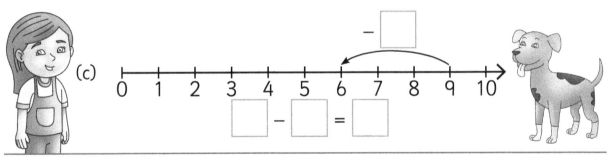

(c)

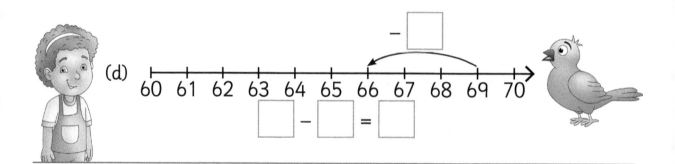

(d)

$$\boxed{} - \boxed{} = \boxed{}$$

2 Look at each table and complete the number sentences.
Write the answers in the boxes.

Tens	Ones
● ● ●	○ ○
	○ ○ ○ ○

$32 + \boxed{} = \boxed{}$

Tens	Ones
● ● ● ● ● ●	○ ○ ○ ⊘ ⊘ ⊘ ⊘

$83 - \boxed{} = \boxed{}$

3 Calculate mentally.

$83 + 4 = \boxed{}$ $93 + 4 = \boxed{}$ $48 - 8 = \boxed{}$ $62 + 3 = \boxed{}$

$39 - 6 = \boxed{}$ $3 + 72 = \boxed{}$ $57 - 5 = \boxed{}$ $48 - 3 = \boxed{}$

$5 + 51 = \boxed{}$ $56 - 2 = \boxed{}$ $32 + 6 = \boxed{}$ $78 - 4 = \boxed{}$

$79 - 8 = \boxed{}$ $88 - 6 = \boxed{}$ $7 + 52 = \boxed{}$ $99 - 5 = \boxed{}$

4 Write the correct number in each box.

(a) 45 $\xrightarrow{+\,3}$ ☐ $\xrightarrow{-\,4}$ ☐ $\xrightarrow{-\,2}$ ☐ $\xrightarrow{+\,7}$ ☐

(b) 68 $\xrightarrow{-\,5}$ ☐ $\xrightarrow{+\,4}$ ☐ $\xrightarrow{+\,2}$ ☐ $\xrightarrow{-\,8}$ ☐

Challenge and extension question

5 Put the nine numbers below into three groups. Then use the numbers to make two addition sentences and two subtraction sentences in the spaces below.

1   2   3   61   62   64   66   67   68

2.3 Adding or subtracting a 1-digit number to or from a 2-digit number (2)

Learning objective Add or subtract a 1-digit number to or from 2-digit numbers

Basic questions

1 Use the number lines to write addition and subtraction calculations. Write the additions and the answers in the correct boxes.

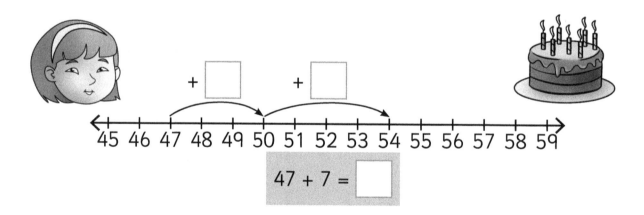

$+\ \square$ $+\ \square$

45 46 47 48 49 50 51 52 53 54 55 56 57 58 59

47 + 7 = $\square$

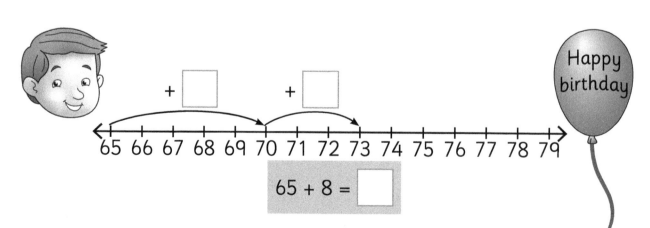

$+\ \square$ $+\ \square$

65 66 67 68 69 70 71 72 73 74 75 76 77 78 79

65 + 8 = $\square$

Happy birthday

2 Solve the problems and write the answers in the boxes. The answer to the last problem in each column should be the same as the answer to the first problem.

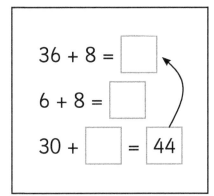

36 + 8 = ☐

6 + 8 = ☐

30 + ☐ = 44

36 + 8 = ☐

36 + ☐ = 40

40 + ☐ = ☐

3 Use your preferred method to work out the calculations.

83 + 4 = ☐

93 + 4 = ☐

48 – 8 = ☐

4 Calculate mentally.

83 + 9 = ☐   39 + 4 = ☐   48 + 8 = ☐   66 + 4 = ☐

39 + 6 = ☐   8 + 72 = ☐   57 + 5 = ☐   3 + 48 = ☐

5 + 58 = ☐   56 + 7 = ☐   9 + 39 = ☐   78 + 4 = ☐

79 + 8 = ☐   88 + 6 = ☐   7 + 54 = ☐   5 + 79 = ☐

5 Write >, < or = in each ◯ to find the answers.

47 + 5 ◯ 54	86 + 7 ◯ 9 + 85	27 + 4 ◯ 25 + 7
89 + 6 ◯ 95	57 + 5 ◯ 6 + 58	5 + 46 ◯ 48 + 6
65 ◯ 54 + 8	8 + 64 ◯ 61 + 9	99 + 1 ◯ 8 + 92

Challenge and extension question

6 There are eight numbers.

Cross through some of the number balloons so that the remaining numbers total 60.

Number sentence: _____

_____

2.4 Adding or subtracting a 1-digit number to or from a 2-digit number (3)

Learning objective Add or subtract a 1-digit number to or from 2-digit numbers

Basic questions

1 Show subtraction using a number line. Write the subtractions and the answers in the correct boxes.

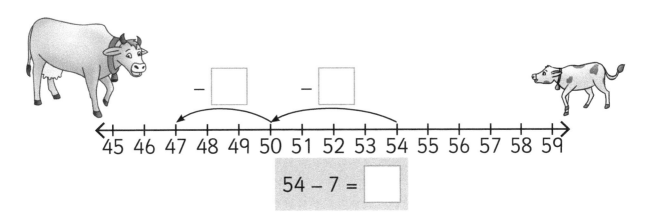

$$54 - 7 = \boxed{}$$

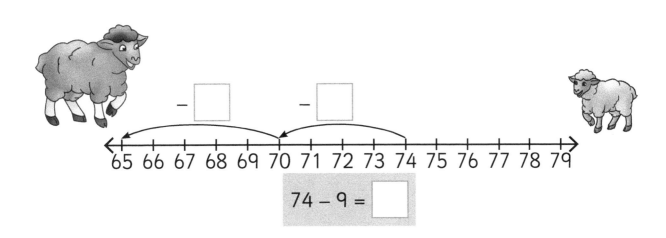

$$74 - 9 = \boxed{}$$

2 Solve the problems and write the answers in the boxes.
The answer to the last problem in each column should be
the same as the answer to the first problem.

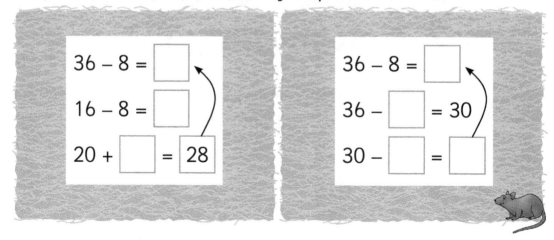

$36 - 8 = \boxed{}$

$16 - 8 = \boxed{}$

$20 + \boxed{} = \boxed{28}$

$36 - 8 = \boxed{}$

$36 - \boxed{} = 30$

$30 - \boxed{} = \boxed{}$

3 Use your preferred method to work out the calculations.

$75 - 9 = \boxed{}$

$42 - 7 = \boxed{}$

$93 - 6 = \boxed{}$

4 Calculate mentally.

$83 - 9 = \boxed{}$ $34 - 5 = \boxed{}$ $84 - 8 = \boxed{}$ $64 - 6 = \boxed{}$

$32 - 8 = \boxed{}$ $80 - 7 = \boxed{}$ $52 - 6 = \boxed{}$ $34 - 8 = \boxed{}$

$85 - 7 = \boxed{}$ $51 - 4 = \boxed{}$ $93 - 9 = \boxed{}$ $87 - 9 = \boxed{}$

$78 - 9 = \boxed{}$ $86 - 8 = \boxed{}$ $54 - 7 = \boxed{}$ $50 - 9 = \boxed{}$

5 Group these number sentences under the correct answer.
Write each number sentence in the correct box.
The first two have been done for you.

~~57 + 8~~   ~~46 – 3~~   73 – 8

5 + 38   59 + 6   50 – 7

63 – 20   70 – 5   25 + 40

52 – 9   51 – 8   2 + 63

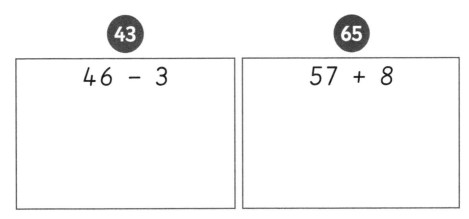

43	65
46 – 3	57 + 8

Challenge and extension question

6 What is the greatest number you can write in each box?

36 + 20 > ☐ + 49   72 – ☐ > 68 – 5

☐ – 6 < 43 + 7   4 + ☐ < 100 – 8

2.5 Adding two 2-digit numbers (1)

 Learning objective Add 2-digit numbers

 Basic questions

1 Look at each table and write the number sentences.

Tens	Ones
●●●	○○
●	○○ ○○

☐ ○ ☐ ○ ☐

Tens	Ones
●●	○○ ○○ ○○
●● ●●	○○

☐ ○ ☐ ○ ☐

2 Solve the problems and write the answers in the boxes. The answer to the last problem in each column should be the same as the answer to the first problem.

42 + 36 = ☐

40 + 30 = ☐

2 + 6 = ☐

70 + ☐ = 78

42 + 36 = ☐

42 + ☐ = ☐

☐ + ☐ = ☐

3 Use your preferred method to work out the calculations.

41 + 26 = ☐

12 + 73 = ☐

62 + 34 = ☐

4 Calculate mentally.

23 + 12 = ☐   36 + 43 = ☐   45 + 31 = ☐   26 + 42 = ☐

32 + 65 = ☐   26 + 62 = ☐   57 + 42 = ☐   53 + 34 = ☐

14 + 85 = ☐   25 + 21 = ☐   14 + 33 = ☐   78 + 21 = ☐

27 + 41 = ☐   43 + 46 = ☐   12 + 54 = ☐   25 + 53 = ☐

5 Mei, Layla and Noah went to a bookshop to buy some books.

Book **A**   Book **B**   Book **C**   Book **D**

Baking

Cars

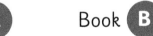
Dinosaurs

Art and Painting

£12   £14   £23   £31

(a) Mei bought Book B and Book D. How much did she pay in total?

Number sentence: _____

Mei paid £ ☐ altogether.

(b) Layla bought Book A and Book C. How much did she spend on these two books?

Number sentence: _____

Layla spent £ ☐ in total.

(c) Noah bought three of the four books. He spent £66 altogether. Which three books did he buy?

Number sentence: _____

Noah bought these books –

Book ☐ , Book ☐ and Book ☐

Challenge and extension question

6 What number does each shape stand for? Write the answers in the boxes.

◯ + ◯ + ◯ = 60	△ + △ + △ + △ = ☆ + ☆

◯ + ◇ + ◇ = 100   ☆ – 30 = 10

◯ = ☐   ◇ = ☐   △ = ☐   ☆ = ☐

2.6 Adding two 2-digit numbers (2)

Learning objective Add 2-digit numbers

Basic questions

1 Look at each table and write down the number sentences.

Tens	Ones
● ● ● ●	○○ ○○ ○○
●	○○ ○○ ○○ ○○

☐ + ☐ = ☐

Tens	Ones
● ●	○○ ○ ○○
● ● ●	○○ ○ ○○

☐ + ☐ = ☐

2 Solve the problems and write the answers in the boxes. The answer to the last problem in each column should be the same as the answer to the first problem.

47 + 38 = ☐

40 + 30 = ☐

7 + 8 = ☐

70 + 15 = ☐

47 + 38 = ☐

47 + ☐ = ☐

☐ + ☐ = ☐

47 + 38 = ☐

47 + 40 = ☐

☐ − ☐ = ☐

3 Use your preferred method to work out the calculations.

| 49 + 26 = ☐ | 14 + 78 = ☐ | 56 + 34 = ☐ |

4 Calculate mentally.

28 + 17 = ☐   35 + 48 = ☐   48 + 34 = ☐   26 + 36 = ☐

39 + 35 = ☐   16 + 77 = ☐   57 + 29 = ☐   57 + 18 = ☐

15 + 85 = ☐   56 + 24 = ☐   17 + 73 = ☐   69 + 24 = ☐

27 + 44 = ☐   49 + 46 = ☐   14 + 58 = ☐   25 + 63 = ☐

5 Draw lines to match the calculations. The first one has been done for you.

36 + 54	16 + 70	68 + 25	16 + 49
47 + 39	36 + 36	4 + 66	48 + 22
68 + 14	52 + 29	47 + 18	25 + 35
53 + 28	45 + 45	53 + 7	53 + 22
19 + 53	24 + 67	39 + 36	88 + 5
75 + 16	7 + 75	28 + 55	57 + 26

Challenge and extension question

6 Think carefully. How can you calculate faster?

Rahim's dad bought boxes of the following different types of fruit:

Type of fruit						
Cost (£)	23	21	12	19	17	8

He paid £ ☐ in total.

2.7 Adding two 2-digit numbers (3)

Learning objective Add 2-digit numbers using a formal written method

Basic questions

1 Do the calculations and write the missing numbers in the boxes. The first one has been done for you.

(a)

$$\begin{array}{r} 4\ 5 \\ +\ 2\ 7 \\ \hline \boxed{7}\ 2 \end{array} \qquad \begin{array}{r} 2\ 8 \\ +\ 4\ 9 \\ \hline \boxed{}\ 7 \end{array}$$

$$\begin{array}{r} 3\ 6 \\ +\ 5\ 4 \\ \hline \boxed{}\ 0 \end{array} \qquad \begin{array}{r} 7\ 3 \\ +\ 1\ 2 \\ \hline \boxed{}\ 5 \end{array}$$

(b)

$$\begin{array}{r} 5\ 8 \\ +\ 3\ 6 \\ \hline \boxed{}\ \boxed{} \end{array} \qquad \begin{array}{r} 6\ 5 \\ +\ 2\ 7 \\ \hline \boxed{}\ \boxed{} \end{array}$$

$$\begin{array}{r} 3\ 9 \\ +\ 1\ 8 \\ \hline \boxed{}\ \boxed{} \end{array} \qquad \begin{array}{r} 4\ 8 \\ +\ 4\ 2 \\ \hline \boxed{}\ \boxed{} \end{array}$$

2 Use the column method to find the answers.

18 + 37 = ☐	56 + 24 = ☐	78 + 15 = ☐	23 + 59 = ☐
25 + 49 = ☐	46 + 47 = ☐	48 + 48 = ☐	72 + 28 = ☐

3 Are these calculations correct? (Put a ✓ for 'yes' and a ✗ for 'no' in the box under each problem and then write the correct calculations below.)

```
    5  4          2  9          3  7          6  4
+   3  8      +   4        +   5  3      +   2  1
---------     ---------     ---------     ---------
    8  2          6  9          9  0          9  5
---------     ---------     ---------     ---------
```

Correct calculations:

4 Complete the table to show the total number of pupils in each year.

Number of pupils doing PE in school

	Year 1	Year 2	Year 3	Year 4	Year 5	Year 6
Boys	29	34	38	55	43	32
Girls	28	27	47	45	39	36
Total						

 Challenge and extension question

5 Find the missing numbers and write them in the boxes.

$$\begin{array}{r} \boxed{}\ 8 \\ +\ 2\ \boxed{} \\ \hline 6\ 5 \end{array} \qquad \begin{array}{r} 2\ \boxed{} \\ +\ \boxed{}\ 4 \\ \hline 7\ 2 \end{array} \qquad \begin{array}{r} \boxed{}\ \boxed{} \\ +\ 5\ 6 \\ \hline 7\ 2 \end{array} \qquad \begin{array}{r} 3\ 9 \\ +\ \boxed{}\ \boxed{} \\ \hline 9\ 4 \end{array}$$

2.8 Adding two 2-digit numbers (4)

Learning objective Add 2-digit numbers using a formal written method

Basic questions

1 Write the answers to the number sentences in the boxes.

36 + 59 = ☐   51 + 49 = ☐   27 + 36 = ☐

47 + 27 = ☐   65 + 27 = ☐   38 + 53 = ☐

15 + 65 = ☐   64 + 31 = ☐   54 + 26 = ☐

17 + 38 = ☐   38 + 44 = ☐   67 + 24 = ☐

63 + 19 = ☐   56 + 25 = ☐   29 + 63 = ☐

63 + 17 = ☐   48 + 35 = ☐   8 + 76 = ☐

2 Use the column method to find the answers.

25 + 63 = ☐	48 + 37 = ☐	9 + 45 = ☐	59 + 21 = ☐
56 + 27 = ☐	46 + 33 = ☐	48 + 48 = ☐	36 + 64 = ☐

3 Are these calculations correct? (Put a ✓ for 'yes' and a ✗ for 'no' in the box under each problem and then write the correct calculations below.)

```
      5           2  9          3  2          6  9
  +   3  8    +   4  7      +   5  3      +   2  1
  ─────────   ─────────     ─────────     ─────────
      8  8       6  6          9  5          9  0
```

Correct calculations:

4 Write the correct numbers in the boxes to complete the addition calculations.

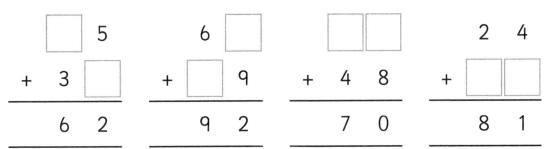

	☐	5		6	☐		☐	☐		2	4
+	3	☐	+	☐	9	+	4	8	+	☐	☐
	6	2		9	2		7	0		8	1

5 Write each number sentence and find the answer.

(a) The two addends are 49 and 25. What is their sum?

_____

The sum of the two addends is ☐.

(b) A is 36 and B is 28 greater than A. What is B?

_____

B is ☐.

Challenge and extension question

6 Try different numbers. What number does each shape stand for?

	○	☆			△	9
+	☆	9		+	△	◇
	8	2			9	△

○ = ☐   ◇ = ☐   △ = ☐   ☆ = ☐

2.9 Subtracting a 2-digit number from a 2-digit number (1)

Learning objective Subtract 2-digit numbers

Basic questions

1 Write the number sentence for each table.

Tens	Ones
•• ••	○○ ⊘⊘ ⊘⊘

☐ ○ ☐ ○ ☐

Tens	Ones
•• • ⊘⊘	○○ ○○ ○⊘ ⊘⊘

☐ ○ ☐ ○ ☐

2 Solve the problems and write the answers in the boxes. The answer to the last problem in each column should be the same as the answer to the first problem.

48 – 32 = ☐

40 – 30 = ☐

8 – 2 = ☐

10 + 6 = ☐

48 – 32 = ☐

48 – 30 = ☐

☐ – 2 = ☐

3 Use your preferred method to work out the calculations.

76 − 45 = ☐	84 − 24 = ☐	59 − 37 = ☐

4 Calculate mentally.

27 − 13 = ☐   35 − 33 = ☐   48 − 31 = ☐   75 − 42 = ☐

65 − 32 = ☐   76 − 32 = ☐   57 − 22 = ☐   55 − 13 = ☐

85 − 14 = ☐   56 − 21 = ☐   76 − 73 = ☐   77 − 21 = ☐

47 − 31 = ☐   46 − 36 = ☐   84 − 54 = ☐   85 − 70 = ☐

5 Stationery shopping.

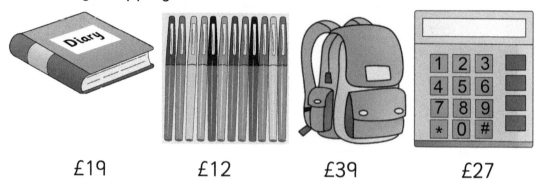

£19   £12   £39   £27

(a) How much more expensive is a schoolbag than a calculator?

Number sentence: _____

A schoolbag is ☐ more expensive than a calculator.

(b) How much cheaper is a diary than a calculator?

Number sentence: _____

A diary is ☐ cheaper than a calculator.

(c) Is the diary cheaper or more expensive than the pens?

Comparison: 19 ◯ 12; so the diary is _____ than the pencils.

Number sentence: _____

The difference in price is ☐ .

Challenge and extension question

6 Fill in the boxes with following numbers to make the number sentence true: 0, 1, 2, 3, 7, 8, 9. You can only use each number once.

☐ + ☐ = ☐ − ☐ = ☐

2.10 Subtracting a 2-digit number from a 2-digit number (2)

Learning objective Subtract 2-digit numbers using a formal written method

Basic questions

1 Write the correct number in each box.

(a)

```
    5  9        6  5        3  9        9  0
 -  3  6     -  2  7     -  1  8     -  4  5
   ┌──┬──┐    ┌──┬──┐    ┌──┬──┐    ┌──┬──┐
   │  │  │    │  │  │    │  │  │    │  │  │
   └──┴──┘    └──┴──┘    └──┴──┘    └──┴──┘
```

(b)

```
    4  7        9  2        7  4        8  1
 -  2  3     -  4  9     -  5  8     -  1  3
   ┌──┐          ┌──┐       ┌──┐       ┌──┐
   │  │ 4        │  │ 3     │  │ 6     │  │ 8
   └──┘          └──┘       └──┘       └──┘
```

2 Use the column method to find the answers.

91 − 37 = ☐	56 − 24 = ☐	70 − 17 = ☐	86 − 59 = ☐
75 − 49 = ☐	86 − 27 = ☐	95 − 45 = ☐	100 − 32 = ☐

3 Are these calculations correct? Put a ✔ for 'yes' and a ✗ for 'no' in the box under each problem and then write the correct calculations below.

```
    5  4          9  1          7  8          8  5
 -  3  8       -  4  7       -  5  3       -  2  5
 ─────────     ─────────     ─────────     ─────────
    2  6          5  6          1  5             9
```

Correct calculations:

4 Complete the table.

Minuend	52		74	51	97	
Subtrahend	28	27		45		39
Difference		33	20		12	43

Challenge and extension question

5 Find the missing number for each box.

```
   ☐  3          7  ☐          ☐  ☐          6  ☐
 -  2  ☐       -  ☐  9       -  5  6       -  ☐  7
 ─────────     ─────────     ─────────     ─────────
    6  5          2  3          3  4             4
```

2.11 Subtracting a 2-digit number from a 2-digit number (3)

 Learning objective Subtract 2-digit numbers using a formal written method

 Basic questions

1 Calculate mentally.

86 − 59 = ☐   51 − 49 = ☐   87 − 35 = ☐   63 − 17 = ☐

47 − 28 = ☐   65 − 27 = ☐   93 − 53 = ☐   48 − 45 = ☐

75 − 65 = ☐   64 − 31 = ☐   54 − 26 = ☐   80 − 72 = ☐

77 − 38 = ☐   83 − 44 = ☐   60 − 24 = ☐   42 − 27 = ☐

63 − 19 = ☐   56 − 25 = ☐   92 − 63 = ☐   83 − 74 = ☐

2 Use the column method to find the answers.

75 – 63 = ☐	83 – 37 = ☐	80 – 42 = ☐	54 – 48 = ☐
56 – 27 = ☐	46 – 33 = ☐	84 – 48 = ☐	100 – 75 = ☐

3 Are these calculations correct? Choose a ✓ for 'yes' or a ✗ for 'no' in the box under each problem and then write the correct calculations below.)

```
    5  0          6  8          8  2          7  1
 -  3  8       -  4  7       -  5  3       -  2  4
 _____     _____     _____     _____
    2  8          1  1          3  9          9  5
 _____     _____     _____     _____
```

Correct calculations:

4 Find the missing number for each box.

$$
\begin{array}{r}
\boxed{}\,5 \\
-\quad 3\ \boxed{} \\
\hline
4\quad 2 \\
\hline
\end{array}
\qquad
\begin{array}{r}
6\ \boxed{} \\
-\ \boxed{}\,9 \\
\hline
1\quad 2 \\
\hline
\end{array}
\qquad
\begin{array}{r}
\boxed{}\boxed{} \\
-\quad 4\quad 5 \\
\hline
3\quad 0 \\
\hline
\end{array}
\qquad
\begin{array}{r}
7\quad 3 \\
-\ \boxed{}\boxed{} \\
\hline
5 \\
\hline
\end{array}
$$

5 Write out the number sentence and find the answer.

(a) If the minuend is 80 and the subtrahend is 25, what is the difference?

Number sentence: _____

The difference between $\boxed{}$ and $\boxed{}$ is $\boxed{}$.

(b) If A is 91 and B is 76, what is the difference between A and B?

Number sentence: _____

The difference between $\boxed{}$ and $\boxed{}$ is $\boxed{}$.

Challenge and extension question

6 Try different numbers. Write the correct numbers in the boxes.

$$
\begin{array}{r}
\triangle\quad \star \\
-\quad \star\quad 8 \\
\hline
3\quad 7 \\
\hline
\end{array}
\qquad\qquad
\begin{array}{r}
\bigcirc\quad 9 \\
-\quad \diamond\quad \bigcirc \\
\hline
1\quad 6 \\
\hline
\end{array}
$$

$\bigcirc = \boxed{}$ $\quad \diamond = \boxed{}$ $\quad \triangle = \boxed{}$ $\quad \star = \boxed{}$

2.12 Adding or subtracting three numbers and mixed operations (1)

Learning objective Add three numbers

Basic questions

1 Calculate mentally.

$6 + 3 + 7 = \boxed{}$  $8 + 6 + 4 = \boxed{}$ $4 + 4 + 3 = \boxed{}$

$6 + 8 + 9 = \boxed{}$ $6 + 6 + 6 = \boxed{}$ $6 + 9 + 5 = \boxed{}$

$4 + 0 + 6 = \boxed{}$ $3 + 9 + 8 = \boxed{}$ $9 + 7 + 8 = \boxed{}$

2 Complete the boxes to solve each addition problem.

(a)

$$\begin{array}{r} 2\ \ 3 \\ +\ 4\ \ 6 \\ \hline \boxed{}\ \boxed{} \\ +\ 1\ \ 8 \\ \hline \boxed{}\ \boxed{} \end{array} \qquad \begin{array}{r} 3\ \ 8 \\ +\ 2\ \ 5 \\ \hline \boxed{}\ \boxed{} \\ +\ 1\ \ 7 \\ \hline \boxed{}\ \boxed{} \end{array} \qquad \begin{array}{r} 4\ \ 9 \\ +\ 1\ \ 6 \\ \hline \boxed{}\ \boxed{} \\ +\ 2\ \ 3 \\ \hline \boxed{}\ \boxed{} \end{array} \qquad \begin{array}{r} 1\ \ 2 \\ +\ 2\ \ 9 \\ \hline \boxed{}\ \boxed{} \\ +\ 4\ \ 6 \\ \hline \boxed{}\ \boxed{} \end{array}$$

(b)

	1 8		2 6		4 9		3 7
	3 5		3 8		2 7		1 9
+	2 4	+	1 9	+	1 5	+	2 4
	☐ 7		☐ 3		☐ 1		☐ 0

3 Use the column method to calculate the answers.

5 + 48 + 37 = ☐	22 + 36 + 19 = ☐	16 + 59 + 24 = ☐
58 + 29 + 8 = ☐	45 + 16 + 15 = ☐	37 + 17 + 27 = ☐

4 Complete the table.

The 2016 Olympic Games Medal Table

	Gold	Silver	Bronze	Total
United States	46	37	38	121
Great Britain & Northern Ireland	27	23	17	
China	26	18	26	
Russia	19	18	19	
Germany	17	10	15	

Challenge and extension question

5 What number does each letter in NUMBER represent?

```
    U  M              2  B
+   U  M          +   5  B
---------         ---------
   N  8              E  0
+      N          +   1  R
---------         ---------
   8  5              9  6
---------         ---------
```

```
N   U   M   B   E   R
☐   ☐   ☐   ☐   ☐   ☐
```

2.13 Adding or subtracting three numbers and mixed operations (2)

 Learning objective Subtract three numbers

 Basic questions

1 Calculate mentally.

$16 - 3 - 7 = \boxed{}$ $18 - 6 - 4 = \boxed{}$ $10 - 5 - 3 = \boxed{}$

$19 - 8 - 9 = \boxed{}$ $16 - 6 - 6 = \boxed{}$ $16 - 9 - 5 = \boxed{}$

$14 - 2 - 6 = \boxed{}$ $12 - 3 - 8 = \boxed{}$ $17 - 9 - 8 = \boxed{}$

2 Complete the boxes to solve each subtraction problem.

```
    9  3          7  8          6  7          8  2
 -  4  6       -  2  5       -  1  9       -  2  9
   ┌──┬──┐       ┌──┬──┐       ┌──┬──┐       ┌──┬──┐
   │  │  │       │  │  │       │  │  │       │  │  │
   └──┴──┘       └──┴──┘       └──┴──┘       └──┴──┘
 -  1  8       -  1  7       -  2  3       -  3  7
   ┌──┬──┐       ┌──┬──┐       ┌──┬──┐       ┌──┬──┐
   │  │  │       │  │  │       │  │  │       │  │  │
   └──┴──┘       └──┴──┘       └──┴──┘       └──┴──┘
```

3 Use the column method to find the answers.

95 – 48 – 37 = ☐	82 – 36 – 19 = ☐	96 – 59 – 24 = ☐
60 – 29 – 8 = ☐	65 – 16 – 35 = ☐	77 – 18 – 28 = ☐

4 Complete the table to show how many of each type of vehicle were left after two checks of the car park.

Number of vehicles

	At the beginning	Drove away (first time)	Drove away (second time)	Remaining
Bus	61	23	27	
Truck	40	18	16	
Car	83	21	38	

5 Fill in the circles with suitable numbers so that the sum of the three numbers on each line is equal to the number in the middle.

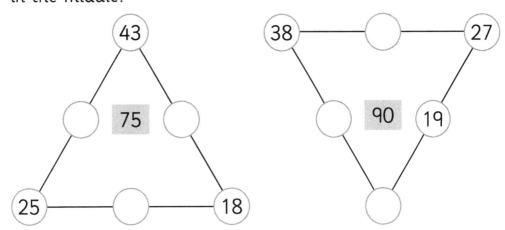

 Challenge and extension question

6 What number does each letter in FUN MATH represent?

$$
\begin{array}{r}
F\ 2 \\
-\ 1\ F \\
\hline
U\ 3 \\
-\ 2\ N \\
\hline
4\ 5 \\
\hline
\end{array}
\qquad
\begin{array}{r}
8\ M \\
-\ A\ 9 \\
\hline
T\ 4 \\
-\ 3\ H \\
\hline
2\ 7 \\
\hline
\end{array}
$$

F	U	N	M	A	T	H
☐	☐	☐	☐	☐	☐	☐

2.14 Adding or subtracting three numbers and mixed operations (3)

Learning objective Add and subtract three numbers

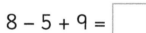

Basic questions

1 Calculate mentally.

$6 - 3 + 7 =$ ☐

$7 + 8 - 9 =$ ☐

$4 + 8 - 3 =$ ☐

$16 - 9 + 8 =$ ☐

$8 - 5 + 9 =$ ☐

$6 + 6 - 7 =$ ☐

$12 - 3 + 8 =$ ☐

$17 - 9 + 4 =$ ☐

2 Complete the boxes to solve each subtraction problem.

9 3	7 8	6 7	8 2
− 4 6	− 2 5	− 1 9	− 2 9
☐ ☐	☐ ☐	☐ ☐	☐ ☐
− 1 8	− 1 7	− 2 3	− 3 7
☐ ☐	☐ ☐	☐ ☐	☐ ☐

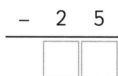
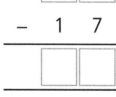

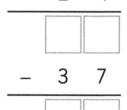

3 Use the column method to find the answers.

95 − 48 + 37 = ☐	28 + 36 − 19 = ☐	93 − 57 + 24 = ☐
46 + 29 − 38 = ☐	65 − 17 + 35 = ☐	72 + 28 − 64 = ☐

4 Are these calculations correct? Put ✓ for 'yes' and ✗ for 'no' in the box under each problem and then write the correct calculations below.)

```
    5  0         2  9         1  2         7  1
 −  3  8      +  4  7      +  5  3      −  2  4
 ─────────    ─────────    ─────────    ─────────
    2  8         7  6         7  5         4  7
 +  4  7      −  1  8      −  3  9      −  5  3
 ─────────    ─────────    ─────────    ─────────
    7  5         9  4         4  6         9  0
```

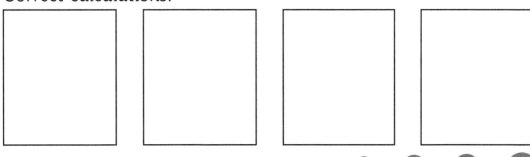

Correct calculations:

Challenge and extension question

5 What number does each letter in FRIENDS represent?

```
      F  0                    E  6
   -  2  R                 +  2  N
   ─────────               ─────────
      5  R                    D  4
   +  I  1                 -  3  S
   ─────────               ─────────
      7  6                    2  9
```

```
   F   R   I   E   N   D   S
  ┌─┐ ┌─┐ ┌─┐ ┌─┐ ┌─┐ ┌─┐ ┌─┐
  └─┘ └─┘ └─┘ └─┘ └─┘ └─┘ └─┘
```

2.15 Problem solving (1)

 Learning objective Addition and subtraction problems

 Basic questions

1 Write the number sentence for each problem and then write the answer in the box.

(a) 35 boys and 27 girls took part in the school concert. How many pupils took part altogether?

Number sentence: _____

☐ pupils took part in the school concert.

(b) Salima made chocolate muffins at the weekend. She made 28 on each of the two days. How many muffins did she make in total?

Number sentence: _____

Salima made ☐ chocolate muffins in total.

(c) Ben is 12 years old and his dad is 38 years old. How many years older than Ben is his dad?

Number sentence: _____

Ben's dad is ☐ years older than Ben.

(d) There are 96 beads in the bead box. 30 of them are green and the rest are red. How many red beads are in the box?

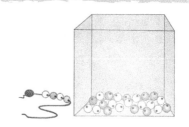

Number sentence: _____

There are ☐ red beads in the bead box.

2 Birthday shopping.

(a) Isaac had £90 to spend for his birthday. He bought a 🏎️. How much did he have left?

Number sentence: _____

Isaac had £ ☐ left.

(b) Isaac also wanted to buy a ⚽ and a 🏈. How much more money did he need?

Number sentence: _____

Isaac needed £ ☐ more.

Challenge and extension question

3 Mo, Tyler and Anya are all reading the same book. Mo has read 72 pages, Tyler has read 75 pages and Anya has read 92 pages. They all plan to finish the whole book.
How many of the following sentences are correct?

(a) Mo has read 3 fewer pages than Tyler.

(b) Tyler has read 17 pages more than Anya.

(c) Mo has read 20 pages fewer than Anya.

Answer: ☐ of the above three sentences are correct.

2.16 Problem solving (2)

Learning objective Addition and subtraction problems

Basic questions

1 Write the number sentences and then calculate. Write the answers in the boxes.

(a) In PE, the pupils took 35 basketballs outside. There were 8 basketballs left inside. How many basketballs were there at the beginning of the lesson?

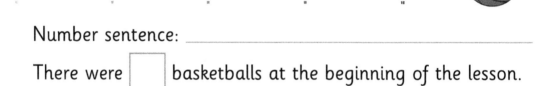

Number sentence: _____

There were ☐ basketballs at the beginning of the lesson.

(b) The fruit shop sold 47 watermelons in the morning and 28 watermelons in the afternoon. How many more watermelons were sold in the morning than in the afternoon?

Number sentence: _____

☐ more watermelons were sold in the morning.

(c) There were 32 fish in the pond. 15 fish swam away and another 26 fish swam into the pond. How many fish are there in the pond now?

Number sentence: _____

Now there are ☐ fish in the pond.

(d) There are 36 fiction books, 18 reading books and 27 non-fiction books in the collection in Ms Patel's classroom. How many books are there altogether?

Number sentence: _____

Altogether there are ☐ books in Ms Patel's classroom.

2 Sports shopping. Write the number sentences and then calculate. Write the answers in the boxes.

£16   £32   £63   £24

(a) How much cheaper is a 🏓 than a 🏸 ?

Number sentence: _____

It is £☐ cheaper.

(b) Joe bought a . He gave the cashier £100.

How much change should he receive?

Number sentence: _____

Joe should receive £ ⬜ in change.

(c) May bought a . Her change was £18.

How much did she give to the cashier?

Number sentence: _____

May gave £ ⬜ to the cashier.

Challenge and extension question

3 Eve has the following coins: 50p, 20p and 10p.
How many different values can she make with these coins? ⬜

What are they?

_____

2.17 Practice and exercise (I)

 Learning objective Addition and subtraction of numbers to 100, including word problems

 Basic questions

1 Calculate mentally.

72 − 8 = ☐   8 + 23 = ☐   79 − 4 = ☐   63 − 7 = ☐

82 − 9 = ☐   9 + 6 = ☐   47 − 8 = ☐   82 + 6 = ☐

47 + 8 = ☐   78 + 6 = ☐   87 − 5 = ☐   74 − 6 = ☐

75 − 9 = ☐   39 + 6 = ☐   4 + 39 = ☐   42 + 4 = ☐

83 − 42 = ☐   47 − 32 = ☐   87 − 6 = ☐   92 − 7 = ☐

2 Use your preferred method to work out the calculations.

68 + 7 = ☐	42 + 34 = ☐	93 − 7 = ☐	86 − 25 = ☐

3 Use the column method to find the answers.

72 − 49 = ☐	9 + 47 = ☐	81 − 53 + 29 = ☐
29 + 46 = ☐	60 − 34 = ☐	92 − 36 − 47 = ☐
53 + 47 = ☐	100 − 38 = ☐	17 + 29 + 38 = ☐

4 Find the missing number for each box.

$$
\begin{array}{cc}
9 & \boxed{} \\
- \ \boxed{} & 2 \\
\hline
3 & 4 \\
\hline
\end{array}
\qquad
\begin{array}{cc}
\boxed{} & 0 \\
- \ 3 & 2 \\
\hline
4 & \boxed{} \\
\hline
\end{array}
\qquad
\begin{array}{cc}
2 & \boxed{} \\
+ \ \boxed{} & 7 \\
\hline
7 & 8 \\
\hline
\end{array}
\qquad
\begin{array}{cc}
\boxed{} & 3 \\
+ \ 4 & \boxed{} \\
\hline
8 & 2 \\
\hline
\end{array}
$$

5 Write the number sentences and then calculate.
Write the answers in the boxes.

(a) Tom bought a dictionary for £36. He gave the cashier a £50 note. How much change did he receive?

Number sentence: _____

Tom received £ ☐ in change.

(b) Thea has read 47 pages of a book. She has 28 more pages to read to reach the end. How many pages does the book have?

Number sentence: _____

Thea's book has ☐ pages.

(c) There are 42 white balloons, which is 7 fewer than the number of red balloons. How many red balloons are there? How many white balloons and red balloons are there altogether?

Number sentences: _____

_____

There are ☐ red balloons.

There are ☐ balloons altogether.

6 Write the missing numbers in the boxes.

(a) A box of chocolates costs £24.

A bag of sweets costs £3.

If Archie buys 1 box of chocolates and gives the cashier £50, he will receive ☐ change.

If he buys 4 bags of sweets and pays the cashier £20, he should receive ☐ change.

If he buys 2 boxes of chocolates and 1 bag of sweets, he needs to pay ☐.

(b) Ciara is reading a 90-page storybook.

She read 29 pages on the first day. This is 8 pages fewer than the number of pages she read on the second day.

Ciara read ☐ pages on the second day.

So far, she has read ☐ pages in total.

Ciara has ☐ pages left to finish the whole book.

Challenge and extension question

7 Mizu and Aaliya have the same number of paper stars.
Mizu gives 20 stars to Hamish and 10 stars to Aaliya.
Now Aaliya has ☐ more stars than Mizu.

Chapter 2 test

1 Calculate mentally.

36 + 50 = ☐   44 + 8 = ☐   42 + 7 = ☐   82 + 13 = ☐

48 + 16 = ☐   75 – 30 = ☐   68 – 8 = ☐   43 – 6 = ☐

23 + 23 = ☐   13 + 82 = ☐   67 – 2 = ☐   14 + 6 = ☐

5 + 65 = ☐   19 + 64 = ☐   41 – 9 = ☐   39 – 8 = ☐

15 – 7 = ☐   30 + 41 = ☐   56 – 22 = ☐   100 – 8 = ☐

36 – ☐ = 30   40 + ☐ = 100   26 = ☐ – 30

57 – 7 = 75 – ☐   ☐ + 30 = 70 – ☐

2 Use your preferred method to work out the calculations.

38 + 57 = ☐   92 – 9 = ☐

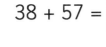

3 Use the column method to find the answers.

48 + 27 = ☐

80 − 56 = ☐

72 − 27 + 55 = ☐

4 Fill in the boxes.

```
   2  ☐
+  ☐  8
―――――――
   6  5
```

```
   ☐  3
−  4  ☐
―――――――
   1  6
```

5 Write >, < or = in each circle.

48 ◯ 54

37 + 18 ◯ 18 + 37

45 + 12 ◯ 86 − 31

10 pounds ◯ 10 pence

32 pence ◯ 32 pounds

88 pence ◯ 99 pence

6 What is the greatest number you can write in each box?

$$20 - \boxed{} > 6 \qquad\qquad \boxed{} + 12 < 16$$

$$\boxed{} - 13 < 45 \qquad\qquad 58 - \boxed{} > 50$$

$$16 - \boxed{} > 2 \qquad\qquad 62 - 40 > \boxed{} + 14$$

7 Write the number sentences and then calculate.

(a) The two addends are 49 and 25. What is the sum?

Number sentence: _____

(b) The minuend is 51 and the difference is 27. What is the subtrahend?

Number sentence: _____

8 Fill in the boxes with suitable numbers.

(a) There are $\boxed{}$ tens and $\boxed{}$ ones in 39.

(b) In a 2-digit number, the tens place is 7 and is 3 greater than the number in the ones place.

The number is $\boxed{}$.

When $\boxed{}$ ones are added, the number is then the smallest 3-digit number.

(c) Counting in twos, the numbers before and after 74 are $\boxed{}$ and $\boxed{}$.

(d) The difference between the smallest 2-digit number and the greatest 2-digit number is ☐.

(e) £1 = ☐ pence.

(f) Making all the different 2-digit numbers with 1, 2 and 9 and then arranging these 2-digit numbers from the greatest to the least, the third number from the left is ☐. You can use each number more than once.

9 Multiple choice questions. (For each question, choose the correct answer and write the letter in the box.)

(a) Tom bought a new schoolbag for the price of 128 ☐.

A. pence   **B.** penny   **C.** pounds   **D.** pound

(b) If you add 4 to one addend and subtract 9 from another addend, the sum is ☐.

A. decreased by 13   **B.** increased by 13

C. increased by 5   **D.** decreased by 5

(c) Ethan and Anila had the same number of sticks.

Ethan gave 10 sticks to Leah and 10 sticks to Anila.

Now Anila has ☐ more sticks than Ethan.

A. 10   **B.** 20   **C.** 30   **D.** 40

10 Comprehensive application.

(a) To celebrate the New Year, the park is decorated with 65 red and blue balloons. 29 of the balloons are blue. How many red balloons are there?

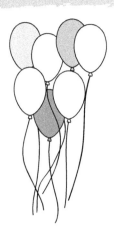

Number sentence: _____

(b) There were 68 crates of milk in a supermarket. After selling 30 crates, another 19 crates were delivered to the supermarket. How many crates of milk are there in the supermarket now?

Number sentence: _____

(c) In a PE lesson, 56 skipping ropes were given out to three classes. Class A and Class B were given 19 ropes each. How many ropes were given to Class C?

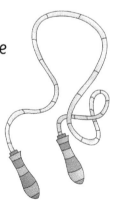

Number sentence: _____

(d) Joe and his friends go to a theme park. Joe has 50 pence to spend on rides. After reading the following price list, Joe decides to choose two different rides.

Pirate ship	25p per ride
Dreamland	35p per ride
Merry-go-round	10p per ride
Space ship	20p per ride
Space walk	30p per ride

(i) Which two of the rides can Joe go on with the money he has?

Show three different options. Draw a 😊 under each of the two rides.

	Pirate ship	Merry-go-round	Space ship	Dreamland	Space walk
Option 1					
Option 2					
Option 3					

(ii) For each of the options you gave above, write a number sentence to show how much money Joe would spend and a number sentence to show how much he will have left.

Option 1: _____

Option 2: _____

Option 3: _____

3.1   Introduction to statistical tables

Learning objective Use tables and tally charts to record and interpret data

Basic questions

1 Count the shapes.

(a) Use a tally chart to record your counting. The first one has been done for you.

△	□	◯	☆	▭	⧄
ⵀ ⵀ ‖‖					

(b) Fill in the table with your results.

Shape	△	□	◯	☆	▭	⧄
Number						

(c) Look at the table and then complete the sentences.

There are ☐ different types of shapes altogether.

The shape that appears the most is _____.

The shape that appears the least is _____.

2 Ethan is the class librarian. He has made a record of the different types of books. Look at his tally chart and complete the statistical table showing the number of each type of book.

comic	ⵀ ‖
science	ⵀ ⵀ
storybook	‖‖‖
cartoon	ⵀ ‖‖‖

Type of book	Number of books
comic	
science	
storybook	
cartoon	

3 In one week, Bella bought a lot of fruit. Use the tally chart to record the number of pieces of fruit she bought. Then complete the statistical table.

Tally chart		Statistical table for number of pieces of fruit	
		Type of fruit	**Number**
🍇		bunch of grapes	
🍎		apple	
🍊		orange	
🍐		pear	
🍌		banana	
🍓		strawberry	

Challenge and extension question

4 A clothes shop received a delivery. The table shows the details.

Type	men's clothes	women's clothes	children's clothes
Number of boxes	10	5	12

(a) Use the table to find the information.

There are ☐ boxes of men's clothes and women's clothes in total.

There are ☐ fewer boxes of women's clothes than children's clothes.

(b) Using the information in the table, what other questions can you ask?

Question: _____

_____

Number sentence: _____

3.2 Pictograms

Learning objective Interpret and construct pictograms

Basic questions

1 The pictogram shows the fruit Dan bought from a shop.

Use the information to complete the sentences.

Dan bought ☐ apples, ☐ oranges and ☐ pears.

He bought ☐ pieces of fruit altogether.

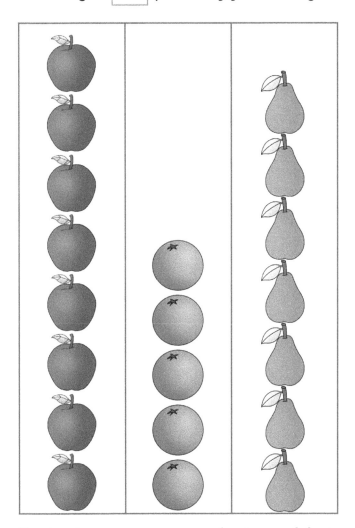

Each picture represents 1 piece of fruit.

2 A class did a survey about each pupil's favourite colour and recorded the results in the pictogram shown below.

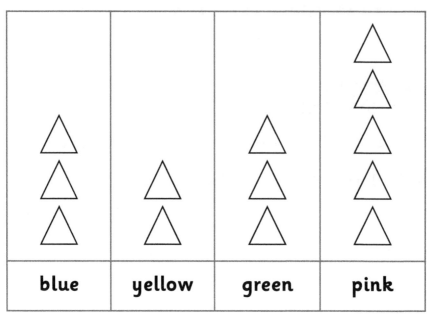

| blue | yellow | green | pink |

Each △ stands for 2 pupils.

Use the information to answer these questions.

(a) What is the favourite colour of the greatest number of pupils in the class? _____

(b) Which colour is least popular? _____

(c) Which colours have been chosen by exactly the same number of pupils? _____

(d) How many pupils chose green as their favourite colour? ☐

(e) How many pupils are there in the class? ☐

3 The table shows the number of pupils in a class who play different sports.

Use the information to make a pictogram. Use one symbol to represent two pupils.

Sports activity

	football	cycling	basketball	swimming
Number of pupils playing	12	10	8	16

football	cycling	basketball	swimming

Challenge and extension question

4 A shop manager used this pictogram to keep a record of the number of computers, of three different brands, sold in a particular week.

Brand A	◯ ◯ ◯ ◯ ◯ ◯ ◯
Brand B	◯ ◯ ◯ ◯ ◯
Brand C	◯ ◯ ◯ ◯ ◯ ◯

Each ◯ stands for 5 computers sold.

(a) Which brand of computers was the most popular that week? How many were sold?

_____

(b) Which brand of computers was the least popular that week? How many were sold?

_____

(c) Do you think the shop should buy in more computers of the most popular brand and stop buying in the least popular brand identified above? Why or why not?

_____

_____

3.3 Block diagrams

 Learning objective Interpret and construct block diagrams

Basic questions

1 The block diagram shows the different types of books in a class. Use it to complete the table below.

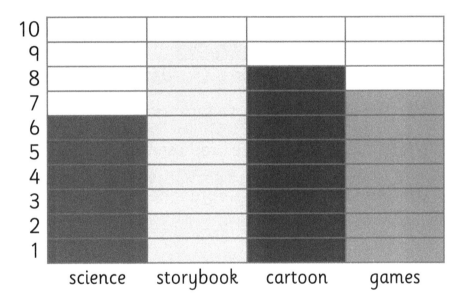

Type of book	science	storybook	cartoon	games
Number of books				

2 The block diagram shows the three categories of the most popular TV programmes voted by pupils in a class.

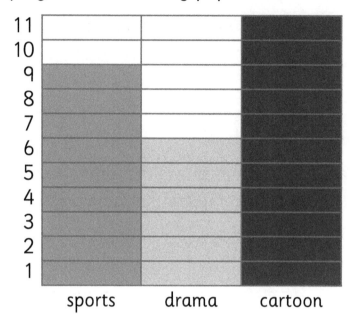

Use the block diagram to find the information.

(a) In the diagram 1 cell represents ☐ pupil(s).

(b) The most popular programme is _____

and the least popular one is _____.

The difference between the two groups is ☐ pupils.

(c) There are ☐ pupils in total in the class.

3 Chloe did a survey of children playing in a playground and produced the following table.

Item	seesaw	slide	swings	sandpit
Number of children	7	8	4	6

Use Chloe's table to make a block diagram.

8				
7				
6				
5				
4				
3				
2				
1				
	seesaw	slide	swings	sandpit

4 Do a survey of 5–10 pupils to find out the number of types of stationery they have in their pencil cases. Complete the table and make a block diagram using the data.

Type	Quantity
pencil	
rubber	
felt-tipped pen	
ruler	

10				
9				
8				
7				
6				
5				
4				
3				
2				
1				

Chapter 3 test

1 The pupils in Class 2R made a tally chart to show the rainfall levels in February.

Level of rainfall	Number of days
Band 1 (low)	~~IIII~~ ~~IIII~~
Band 2 (moderate)	~~IIII~~ ~~IIII~~ ~~IIII~~ I
Band 3 (high)	~~IIII~~
Band 4 (very high)	

(a) There are ☐ days in the month with a low level of rainfall.

(b) The level of rainfall with the greatest number of days is in Band ☐. There were ☐ such days.

(c) There are ☐ days in the month with a very high rainfall level.

2 Malik sorted out all the books in the reading corner of his classroom and made a block diagram to show the information.

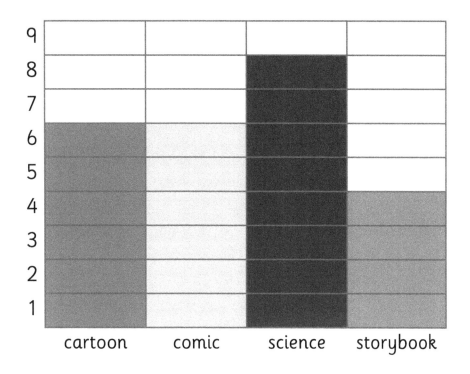

(a) In the diagram, 1 cell stands for ▢ book(s).

(b) In the reading corner, the greatest number of books are

_____ books.

There are ▢ books of this type.

The smallest number of books are _____ books.

There are ▢ such books.

(c) There are ▢ books altogether in the reading corner.

3 At the beginning of the new school term, these new books are bought for the reading corner.

Type of book	cartoon	comic	science	storybook
Number of books	4	3	2	5

Use Malik's block diagram and the data about the new books to make a new table.

Type of book	cartoon	comic	science	storybook
Number of books				

Now make a new block diagram using your table.

10				
9				
8				
7				
6				
5				
4				
3				
2				
1				
	cartoon	comic	science	storybook

4 Amy did a survey in class and asked pupils about their favourite flowers. The pictogram below shows some of the results.

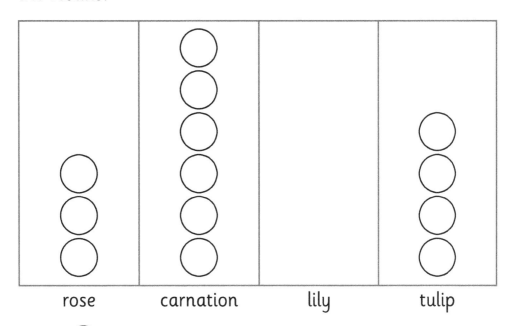

| rose | carnation | lily | tulip |

Each ◯ stands for 2 pupils.

(a) Complete the pictogram: the number of pupils who chose lily as their favourite flower is half the number of pupils who chose tulip.

(b) Using the diagram above, fill in the table.

Favourite flower	rose	carnation	lily	tulip
Number of pupils				

(c) The number of pupils who prefer tulips is ⬚ more than the number of pupils who prefer roses.

The number of pupils who prefer roses is _____
of the number of the pupils who prefer carnations.

Chapter 4   Consolidation and enhancement

4.1   The 100 square (II)

Learning objective Explore number patterns on the 100 square

Basic questions

1 Complete the number patterns.

(a) 76 ☐ 78 ☐ ☐ 81 ☐ ☐ ☐

(b) 100 ☐ ☐ 85 ☐ 75 ☐ ☐ ☐

(c) 56 ☐ ☐ 62 ☐ 66 ☐ ☐ ☐

(d) ☐ 89 ☐ 85 ☐ ☐ 79 ☐ ☐

2 Write the correct numbers in the table based on the 100 square.

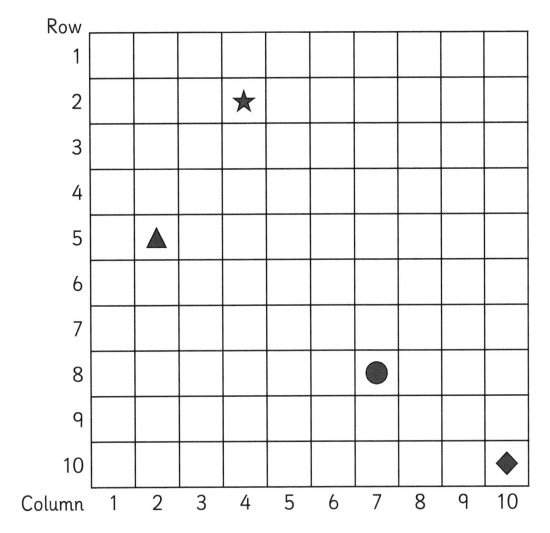

Shape	Row	Column	Number
★	2	4	14
▲			
●			
◆			

3 Use a 100 square to complete the table and fill in the boxes.

Row number	2	8	6	4	10	9
Column number	7	6	7	5	4	10
Number						

The number 4 rows above 75 is ☐.

The number 2 rows below 36 is ☐.

The number 7 columns right from 62 is ☐.

The number 5 columns left from 98 is ☐.

The number 2 rows above 81 is ☐.

The number 3 columns right from 53 is ☐.

4 Answer the following questions based on this 100 square.

1	2	3	4	5	6	7	8	9	10
				※					
									◎
		■							
▲									
						●			
								★	

(a) Find the number each shape below stands for.

※ = ☐   ■ = ☐

● = ☐   ★ = ☐

◎ = ☐   ▲ = ☐

(b) 77 is in the _____ row and _____ column.

(c) Find the position for 88 and draw a ☺ in it.

(d) The number three rows below 64 is ☐ .

The number four columns left from 80 is ☐ .

The number four rows below 58 is ☐ .

Challenge and extension question

5 Work out the calculations. Can you make sense of your answers?

1	2	3	4	5	6	7	8	9	10
11	12	13	14	15	16	17	18	19	20
21	22	23	24	25	26	27	28	29	30
31	32	33	34	35	36	37	38	39	40
41	42	43	44	45	46	47	48	49	50
51	52	53	54	55	56	57	58	59	60
61	62	63	64	65	66	67	68	69	70
71	72	73	74	75	76	77	78	79	80
81	82	83	84	85	86	87	88	89	90
91	92	93	94	95	96	97	98	99	100

11	12	13
21	22	23
31	32	33

4	5	6
14	15	16
24	25	26

17	18	19
27	28	29
37	38	39

21 + 22 + 23 = ☐   14 + 15 + 16 = ☐   27 + 28 + 29 = ☐

12 + 22 + 32 = ☐   5 + 15 + 25 = ☐   18 + 28 + 38 = ☐

11 + 22 + 33 = ☐   4 + 15 + 26 = ☐   17 + 28 + 39 = ☐

13 + 22 + 31 = ☐   6 + 15 + 24 = ☐   19 + 28 + 37 = ☐

4.2 2-digit number addition and subtraction revision (1)

Learning objective Addition and subtraction of numbers to 100, including word problems

Basic questions

1 Group these additions and use the table to record your results.

44 + 27

5 + 76

18 + 5

18 + 20

62 + 16

34 + 43

23 + 5

91 + 4

8 + 21

58 + 18

	Adding ones and a 2-digit number	Adding two 2-digit numbers
Addition without regrouping		
Addition with regrouping		

2 Use your preferred method to work out the calculations.

41 + 26 = ☐   23 + 35 = ☐   23 + 46 = ☐

35 + 61 = ☐   33 + 22 = ☐   54 + 45 = ☐

 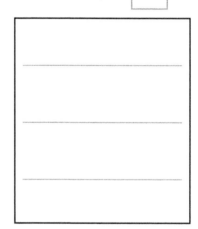

23 + 57 = ☐   53 + 39 = ☐   28 + 46 = ☐

3 Fill in the boxes.

(a) A number has the digit 1 in the tens place and 9 in the ones place. The number is ☐.

(b) 5 tens and 3 ones equal ☐.

24 is made up of ☐ ones.

(c) Reading from the right, the first digit is 5 and the second digit is 3.

This number is ☐.

The number 2 tens greater than this number is ☐.

(d) Counting in fives, the numbers before and after 65 are ☐ and ☐.

(e) The even numbers before and after 18 are ☐ and ☐.

The odd numbers before and after 69 are ☐ and ☐.

4 Answer these word problems.

(a) There are 35 white sheep and 42 black sheep. How many sheep are there altogether? Draw a bar model and write the number sentence.

Number sentence: _____

There are ☐ sheep altogether.

(b) There are 30 cars left after 14 cars drive away from a car park. How many cars were there at first? Draw a bar model and write the number sentence.

Number sentence: _____

There were ⬜ cars in the car park at first.

(c) There are 18 people in the choir. There are 15 more people in the dance group than in the choir. How many people are there in the dance group? Look at the bar model shown below and then write the number sentence.

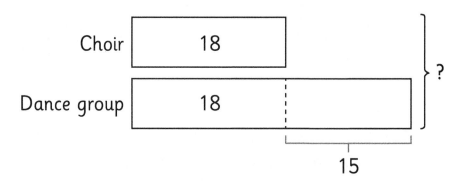

Number sentence: _____

There are ⬜ people in the dance group.

(d) Year 2 pupils are helping to spring-clean their classrooms. 16 pupils are sweeping the floors, 12 pupils are cleaning the whiteboards, the walls and the doors. Another 14 pupils are putting the desks and chairs in order. How many pupils are there helping? Draw a bar model and then write the number sentence.

Number sentence: _____

[ ] pupils are helping to clean their classrooms.

Challenge and extension question

5 Fill in the missing numbers.

+ →		
28	7	
16	15	

+ →		
35	27	
8	14	

4.3   2-digit number addition and subtraction revision (2)

Learning objective   Addition of numbers to 100, including word problems

Basic questions

1 Use your preferred method to work out the calculations.

74 + 15 = ☐

26 + 37 = ☐

57 + 38 = ☐

44 + 48 = ☐

16 + 77 = ☐

54 + 45 = ☐

2 Use the column method to work out the calculations.

22 + 35 = ☐

56 + 19 = ☐

27 + 38 = ☐

69 + 22 = ☐

28 + 55 = ☐

64 + 27 = ☐

72 + 28 = ☐

55 + 39 = ☐

3 Fill in the table.

Addend	32	28	49	18	17
Addend	46	71	27	28	50
Sum					

Addend	41	46	63	59	66
Addend	59	25	18	28	26
Sum					

4 Find the answers to these word problems.

(a) After 25 people get off the bus at a bus stop, there are 27 people left on the bus. How many people were on the bus before it got to the bus stop? Draw a bar model and then write the number sentence.

Number sentence: _____

☐ people were originally on the bus.

(b) There are 36 motorbikes. There are as many bicycles as motorbikes. How many motorbikes and bicycles are there altogether? Draw a bar model and then write the number sentence.

Number sentence: _____

There are ☐ motorbikes and bicycles altogether.

(c) Samira has 28 red beads and 16 yellow beads. How many beads does Samira have in all? Draw a bar model and then write the number sentence.

Number sentence: _____

Samira has ☐ beads altogether.

(d) Keir has £54. He has £16 less than Lydia. How much does Lydia have? Draw a bar model and then write the number sentence.

Number sentence: _____

Lydia has £ ☐ .

(e) Thea went shopping with £50. She wanted to buy a doll for £26 and a box of jigsaw puzzles for £35. Did she have enough money to buy these two toys? Draw a bar model and then write the number sentence.

Number sentence: _____

Answer: _____

Challenge and extension question

5 Fill in the boxes with suitable numbers.

```
  ☐ 4          1 ☐          ☐ 5
+ 2 ☐        + ☐ 8        + 2 ☐
-----        -----        -----
  7 2          9 0          5 2
```

```
  ☐ ☐          2 7          ☐ 9
+ 4 ☐        + 6 ☐        + 3 ☐
-----        -----        -----
  7 6        ☐ 0          8 4
```

4.4 2-digit number addition and subtraction revision (3)

 Learning objective Addition and subtraction of numbers to 100

 Basic questions

1 Group these subtractions and use the table to record your results.

84 – 59   31 – 13   64 – 8

77 – 44   96 – 3

28 – 8

21 – 9   48 – 9   56 – 35   55 – 19

	Subtracting ones and a 2-digit number	Subtracting two 2-digit numbers
Subtraction without regrouping		
Subtraction with regrouping		

2 Choose your preferred method to work out the calculations.

87 − 5 = ☐   87 − 16 = ☐   87 − 40 = ☐

59 − 16 = ☐   33 − 11 = ☐   68 − 35 = ☐

83 − 47 = ☐   63 − 36 = ☐   74 − 59 = ☐

3 Complete the tables.

Addend	42	20			17
Addend	39		24	28	50
Sum		78	87	69	

Minuend	42		87		66
Subtrahend	6	20		28	26
Difference		78	9	28	

4 What is the greatest number you can write in each box?

 6 + ☐ < 19   8 + ☐ < 40   28 > ☐ + 9

 34 − ☐ > 6   ☐ − 16 < 26   5 < 73 − ☐

5 Write the addition or subtraction sentence and then complete the answer to each question.

(a) What number is 27 more than 48?

☐ ◯ ☐ ◯ ☐

☐ is 27 more than 48.

(b) The minuend is 86 and the subtrahend is 53. What is the difference?

☐ ◯ ☐ ◯ ☐

The difference between 86 and 53 is ☐ .

(c) The sum of two numbers is 79. One of the addends is 55. What is the other addend?

☐ ◯ ☐ ◯ ☐

The other addend is ☐ .

(d) After adding 18 to a number, the sum is 99. What is the number?

☐ ◯ ☐ ◯ ☐

The number is ☐ .

 Challenge and extension question

6 Fill in the missing numbers.

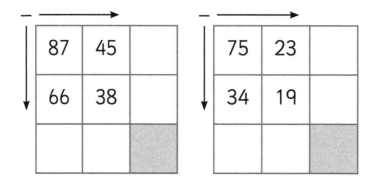

4.5 2-digit number addition and subtraction revision (4)

Learning objective Addition and subtraction of numbers to 100, including word problems

Basic questions

1 Use your preferred method to work out the calculations.

57 − 15 = ☐

53 − 18 = ☐

71 − 58 = ☐

46 − 27 = ☐

76 − 58 = ☐

53 − 35 = ☐

2 Use the column method to work out the calculations.

62 – 25 = ☐

53 – 16 = ☐

63 – 17 = ☐

74 – 49 = ☐

83 – 58 = ☐

91 – 34 = ☐

90 – 22 = ☐

72 – 46 = ☐

92 – 48 – 37 = ☐

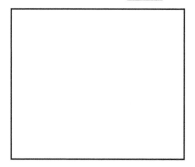  

84 – 45 – 27 = ☐

100 – 46 – 44 = ☐

3 Answer these word problems.

(a) A shop sold 35 washing machines in a day. 18 of them were sold in the morning. How many were sold in the afternoon? Draw a bar model and write the number sentence.

Number sentence: _____

☐ washing machines were sold in the afternoon.

(b) A stationery shop sold 36 pencils and had 28 pencils left. How many pencils were there in the shop at first? Draw a bar model and write the number sentence.

Number sentence: _____

There were ☐ pencils in the shop at first.

(c) There are 27 red balls and 40 white balls in a sports shop. How many more white balls are there than red balls? Draw a bar model and write the number sentence.

Number sentence: _____

There are ☐ more white balls than red balls.

(d) In a shop there are 58 white balloons, which is 19 more than red balloons. How many red balloons are there? Draw a bar model and write the number sentence.

Number sentence: _____

There are ☐ red balloons.

(e) Theo has 80 mental maths problems to complete. On the first day, he did 37 of them. On the second day, he did 25. How many mental sums has he done? How many does he still need to do? Draw a bar model and write the number sentences.

Number sentence: _____

Theo has completed ☐ sums.

He still has ☐ left to do.

(f) Maya bought a football for £35 and a pair of table tennis bats for £28 from a sports shop. She gave the cashier £100. How much change did she get? Draw a bar model and write the number sentence.

£35

Number sentence: _____

Maya got £ ☐ change.

4 Fill in the boxes with suitable numbers.

```
  ☐ 3        5 ☐        6 ☐        ☐ 7
+ 5 ☐      - ☐ 6      + ☐ 6      - 2 ☐
─────      ─────      ─────      ─────
  8 ☐          9        9 3        3 8
```

```
  ☐ 6        3 2        7 1        8 ☐
+ 3 ☐      + ☐ 8      - ☐ ☐      - ☐ 8
─────      ─────      ─────      ─────
    4        7 ☐        6 ☐        1 4
```

119

4.6 Changing the order of numbers in addition (1)

Learning objective Use the commutative law to add numbers in any order

Basic questions

1 Work out the calculations. What do you notice?

23 + 45 = ☐   67 + 8 = ☐   56 + 34 = ☐   49 + 18 = ☐

45 + 23 = ☐   8 + 67 = ☐   34 + 56 = ☐   18 + 49 = ☐

37 + 5 = 5 + ☐   29 + 43 = ☐ + 29   87 + ☐ = 12 + 87

68 + 11 = 11 + ☐   55 + 16 = ☐ + 55   77 + ☐ = 23 + 77

2 Look carefully and then calculate.

48 + 13 = ☐   68 + 5 = ☐   20 + 32 = ☐   7 + 7 = ☐

48 + 14 = ☐   6 + 57 = ☐   23 + 29 = ☐   17 + 57 = ☐

48 + 15 = ☐   46 + 7 = ☐   26 + 26 = ☐   27 + 47 = ☐

16 + 48 = ☐   8 + 35 = ☐   29 + 23 = ☐   37 + 37 = ☐

17 + 48 = ☐   24 + 9 = ☐   32 + 20 = ☐   47 + 37 = ☐

18 + 48 = ☐   10 + 13 = ☐   35 + 17 = ☐   47 + 47 = ☐

3 Calculate with reasoning.

8 + 7 = ☐   65 + 6 = ☐   33 + 19 = ☐

7 + 18 = ☐   6 + 55 = ☐   19 + 13 = ☐

28 + 17 = ☐   55 + 36 = ☐   13 + 79 = ☐

27 + 48 = ☐   36 + 5 = ☐   79 + 3 = ☐

4 Choose three numbers from each group. Use the numbers to make two addition sentences and two subtraction sentences.

17   9   26   8   20   46   76   26

_____   _____

_____   _____

_____   _____

_____   _____

5 Use the column method to work out the answers.

38 + 47 = ☐   54 + 29 = ☐

90 − 42 = ☐   85 − 38 = ☐

6 Answer these word problems.

(a) There were 62 books in Mr Wilson's classroom. Eva borrowed 29 of them. How many books were left? Draw a bar model and write the number sentence.

Number sentence: _____

There were ☐ books left in the classroom.

(b) There were 43 rabbits on the grass. Another 17 rabbits joined them. How many rabbits were there on the grass altogether? Draw a bar model and write the number sentence.

Number sentence: _____

There were ☐ rabbits altogether.

(c) Adam spent his Saturday and Sunday reading his storybook. He read 25 pages on each day. How many pages did he read in the two days? Draw a bar model and write the number sentence.

Number sentence: _____

Adam read ☐ pages in two days.

(e) Finn and Bella had a jumping contest. They did 90 jumps altogether. Bella did 58 jumps. How many jumps did Finn do? Draw a bar model and write the number sentence.

Number sentence: _____

Finn did ☐ jumps.

(f) 65 boats were docked in the harbour. After some boats sailed away, there were 36 boats remaining. How many boats sailed away from the harbour? Draw a bar model and write the number sentence.

Number sentence: _____

☐ boats sailed away from the harbour.

7 Think carefully and then fill in the boxes.

First number + Second number = Second number + ☐

B + D = ☐ + B

◆ + ▲ = ☐ + ◆

a + c = c + ☐

4.7   Changing the order of numbers in addition (2)

 Learning objective   Use patterns of results to add 2-digit numbers

 Basic questions

1 Look carefully and then calculate.

57 + 6 = ☐   40 + 17 = ☐   35 + 55 = ☐

10 + 55 = ☐   19 + 37 = ☐   50 + 39 = ☐

53 + 14 = ☐   34 + 21 = ☐   43 + 45 = ☐

18 + 51 = ☐   31 + 23 = ☐   40 + 47 = ☐

22 + 49 = ☐   25 + 28 = ☐   51 + 35 = ☐

47 + 26 = ☐   25 + 27 = ☐   55 + 30 = ☐

2 Calculate with reasoning.

42 + 13 = ☐   68 + 15 = ☐   49 + 12 = ☐

39 + 16 = ☐   70 + 15 = ☐   13 + 49 = ☐

36 + 19 = ☐   ☐ + 15 = ☐   ☐ + 14 = ☐

33 + 22 = ☐   ☐ + ☐ = ☐   ☐ + 49 = ☐

30 + 25 = ☐   ☐ + ☐ = ☐   49 + ☐ = ☐

3 Write >, < or = in each ◯.

5 + 12 ◯ 29   40 + 16 ◯ 41   91 ◯ 65 + 26

10 + 13 ◯ 29   14 + 37 ◯ 41   91 ◯ 24 + 67

14 + 15 ◯ 29   34 + 12 ◯ 41   91 ◯ 22 + 69

20 + 15 ◯ 29   31 + 10 ◯ 41   91 ◯ 20 + 71

25 + 16 ◯ 29   8 + 28 ◯ 41   91 ◯ 18 + 73

17 + 30 ◯ 29   25 + 6 ◯ 41   91 ◯ 75 + 16

4 Choose three numbers from each group. Use the numbers to make two addition sentences and two subtraction sentences.

25   16   41   9   │   10   36   72   46

───────────────────────   │   ───────────────────────

───────────────────────   │   ───────────────────────

───────────────────────   │   ───────────────────────

5 The sums in each column are in the wrong order. Complete the sums. Then write numbers in the triangles to show the correct order to reach the target number. Part of the first column is already filled in.

△1 8 + 6 = 14   △ 3 + 29 = ☐   △ 48 + 8 = ☐

△5 88 + 12 = ☐   △ 63 + 9 = ☐   △ 72 + 8 = ☐

△4 60 + 28 = ☐   △ 32 + 14 = ☐   △ 80 + 8 = ☐

△2 14 + 27 = 41   △ 72 + 18 = ☐   △ 64 + 8 = ☐

△3 41 + 19 = ☐   △ 46 + 17 = ☐   △ 56 + 8 = ☐

Target ⟩ ☆100☆   Target ⟩ ☆90☆   Target ⟩ ☆88☆

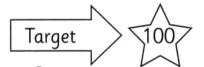

Challenge and extension question

6 Work out these word problems.

(a) A toy shop sold 36 toys in February and 54 toys in March. How many toys were sold in the two months? Draw a bar model and write the number sentence.

Number sentence: _____

☐ toys were sold in the two months.

(b) Conor bought a notebook for £18. He gave the cashier £50. How much change did he get? Draw a bar model and write the number sentence.

Number sentence: _____

Conor got £ ☐ in change.

(c) A bakery has made 65 muffins and 50 doughnuts. How many more doughnuts does it need to make to match the number of muffins? Draw a bar model and write the number sentence.

Number sentence: _____

The bakery needs to make ☐ more doughnuts to match the number of muffins.

(d) A rope was 100 metres long. 54 metres was cut off, and then another 33 metres was cut off. How many metres long is the rope after both cuts?

Number sentence: _____

The rope is ☐ metres long after both cuts.

4.8 Practice and exercise (II)

 Learning objective Solve addition and subtraction problems, interpreting and comparing data

 Basic questions

1 The tally chart shows the types of transport that Year 2 pupils use to travel to school.

Use the chart to complete the table and then make a pictogram. Use one symbol to represent 5 pupils.

walk	ЖЖ ЖЖ ЖЖ ЖЖ ЖЖ ЖЖ
car	ЖЖ ЖЖ ЖЖ ЖЖ ЖЖ ЖЖ ЖЖ ЖЖ ЖЖ ЖЖ
bus	ЖЖ ЖЖ ЖЖ ЖЖ

Transport	Number of pupils
walk	
car	
bus	

walk	
car	
bus	

Now use the information to answer these questions.

Which type of transport is used the most by the Year 2 pupils?

Answer: _____

_____

How many pupils go to school by car or bus?

Answer: _____

_____

2 Work out these word problems.

(a) There are 34 pupils in Class A, 29 pupils in Class B and 36 pupils in Class C.

(i) How many fewer pupils are there in Class B than in Class C?

Answer: _____

There are ⬚ fewer pupils in Class B.

(ii) How many pupils are there in the three classes in total?

Answer: _____

There are ⬚ pupils altogether.

(iii) If one school bus can seat 50 pupils, are two school buses enough to seat all the pupils?

Answer: _____

_____

(b) This is the cost for each pupil of the school trip to the fair:

Admission: £5

Carousel: £4

Roller coaster: £3

Water slide: £6

Coach: £15

How much does each pupil need to pay for the trip?

Answer: _____

_____

(c) Shopping.

£18   £5   £12   £37   £23

(i) Bo bought an alarm clock and a football.
How much did it cost him in total?

Answer: _____

It cost Bo £ ☐ in total.

(ii) Ava wanted to buy a toy train. She gave the cashier a £50 note. How much change did she get?

Answer: _____

Ava got £ ☐ in change.

(iii) Chandra bought two items for £60 exactly. The items she bought cost ☐ and ☐.

(iv) If you were allowed to buy two of the above items, which two would you choose?

How much would they cost?

Answer: I would choose _____ and _____.

They cost £ ☐.

Challenge and extension question

3 How to save?

Tickets for children

A. Individual: £2 per person

B. Group: £15 for 10 people

34 school children are visiting a theatre.

Can you work out the best option for saving money on their tickets?

☐ group tickets and ☐ individual tickets.

It will cost £ ☐ in all.

Number sentence: _____

Chapter 4 test

1 Calculate mentally.

8 + 26 = ☐   12 + 28 = ☐

5 + 25 + 30 = ☐   51 + ☐ = 60

44 – 7 = ☐   28 + 52 = ☐

100 – 50 – 9 = ☐   66 – ☐ = 40

61 – 16 = ☐   48 + 32 = ☐

49 + 1 – 40 = ☐   30 + ☐ = 75 – 7

30 + 70 = ☐   80 – 12 = ☐

37 – 3 + 7 = ☐

2 15 Year 2 pupils took part in a dance competition in three groups. Group A had 4 pupils and Group B had 6 pupils.

Complete the table to show the number of pupils in each group and then make a block diagram with this information.

Group	A	B	C
Number of pupils			

7
6
5
4
3
2
1

Group A   Group B   Group C

3 Choose your preferred method to calculate.

$36 + 64 =$ ☐

$84 - 35 =$ ☐

4 Use the column method to calculate.

$37 + 49 =$ ☐

$76 - 67 =$ ☐

$8 + 24 + 62 =$ ☐

$92 - 36 + 24 =$ ☐

$60 - 37 - 15 =$ ☐

5 Write >, < or = in each ◯ and write a number in each ▢.

26 + 5 ◯ 35	31 – 9 ◯ 22
76 – 30 ◯ 10 + 46	£2 and 80p ◯ 28p
▢ + 21 < 43	48 – ▢ > 20
40 m + 16 m = ▢ m	72 cm – 23 cm = ▢ cm

6 Draw circles to group and then fill in the boxes
(for example 3 fives).

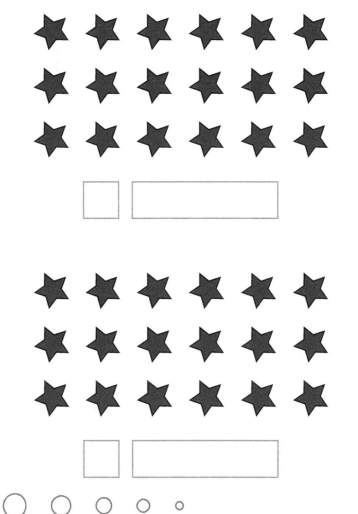

▢ ▭

▢ ▭

7 Work out these word problems.

(a) There are 30 pupils in Class A, 29 pupils in Class B and 32 pupils in Class C.

They are all going on a field trip with 6 teachers.

Each coach can seat 50 people. Will 2 coaches be enough for the trip?

Draw a 😊 in the ☐ next to the correct answer.

A. They will be enough. ☐

B. They will not be enough. ☐

(b) Andrew has £25 and he plans to spend all his money. How can he use his money?

Put a ✓ against the rides he plans to go on, for example Plan 1. Then write the number sentences below.

Attraction	Price	Plan 1	Plan 2	Plan 3	Plan 4
Water slide	£20 per ride per person				
Roller coaster	£15 per ride per person				
Teacups	£5 per ride per person	✓			
Bumper cars	£10 per ride per person	✓ ✓			
Pirate ship	£5 per ride per person				

(i) Plan (1): £5 + £10 + £10 = £25

(ii) Plan (2): _____

(iii) Plan (3): _____

(iv) Plan (4): _____

8 Use the 100 square to answer the questions.

1	2	3	4	5	6	7	8	9	10
11								19	
	22						28		
		33				37			
			44		46			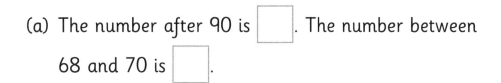	
				55					
			64		66				
		73				77			
	82		★				88		
91								99	

(a) The number after 90 is ☐ . The number between 68 and 70 is ☐ .

(b)  ★ is in the ☐ row and ☐ column.

(c) The number four rows below 28 is ☐ .

(d) ▲ stands for number ☐ and the number 7 columns left from it is ☐ .

(e) The digits in the _____ places of all numbers in the 8th column is 8.

Notes